A FOCUS ON
ENVIRONMENT THROUGH
THIRUKKURAL

Holistic Approach in Thirukkural
Towards Environmental Challenge

S. KALAIVANAN

notionpress
.com

INDIA · SINGAPORE · MALAYSIA

Notion Press

No.8, 3rd Cross Street,
CIT Colony, Mylapore, Chennai,
Tamil Nadu – 600004

First Published by Notion Press 2020
Copyright © S. Kalaivanan 2020
All Rights Reserved.

ISBN 978-1-64892-663-1

Dedication

I dedicate this to my parents R. Sundaram and S. Danam, because of whom I am here, my uncles A. Kaliyaperumal and A. Sarangan who sowed the concept of writing a book in my childhood days and finally Professor. A. Jagadeesan, Former HOD – Tamil, V.O.C College, Tuticorin who inspired me and insisted me to read one Tirukkural daily.

CONTENTS

PART – 2

Prof. Dr.M.SANTHANAMUTHU, Ph.D.,
M.E (Petrochem)., B.E (Chem)., PGDPM.,
PGDBA., FIE., MIIChE., MISTE., C.Engr (I).

Visiting Professor

Formerly

Chief Manager, SPIC Limited
Director, Bharathidasan University
Dean, Anna University, Tiruchirapalli
Controller, Periyar Maniammai University

Residence : 5/283, Kanagavel Nagar, Third Street, Athikulam, K.Pudur, Madurai - 625 007, Tamilnadu, INDIA.
Email : santhanamuthu@yahoo.co.in   Cell : 94433 88358

PREFACE

Everyone is aware of Environment degradation and its global impact. Declining interest in safe keeping environs by a section of society membership erupt unexpected problem. I have read this book and understand that Mr.S.Kalaivanan is involving in a big way bringing to bear on his harness knowledge in Thirukkural to make reforms in present environment difficult situation for the benefit of society. The author attempts to provide remedial action to abate pollution through his findings from Thirukkural in a lucid language. Scientific narration is furnished on how natural process drift ecosystem while adopting new technology. Cause and effect of factors in today life style is analysed with reference to Thriukkural statement. Younger generation and lesser - privileged citizens must play active role to combat environment deterioration. To my mind this is the need of the hour. I like author and book.

With all good wishes for your noble endeavour

η.

Professor Dr.M.Santhanamuthu

INTRODUCTION

I am a strong believer of Tirukkural and its concepts.

Having read and understood all the 1330 kurals I was always fascinated by the in-depth analysis of all aspects of life it provided.

When I completed my Master degree in Environmental science, I realised I could relate many modern concepts in Ecology with Tirukkural.

I was amazed by the sheer genius of Tiruvalluvar who had foreseen many of the modern era concepts before 2000 years.

Propelled by the urge to convey this indispensable information to as many people as possible, I set out to write this book explaining Environmental science on the backdrop of Tirukkural.

This book reveals numerous facts about our Environment, solar system, threats to our atmosphere, biosphere, hydrosphere and lithosphere. Especially Part 2 of this book contains many mind boggling facts about Environment.

PART 1

SUSTAINABLE DEVELOPMENT

செயற்கை அறிந்த கடைத்தும் உலகத்து
இயற்கை அறிந்து செயல்.

Though knowing all that books can educate, the 'truest tact
is To follow common sense of men in act.

நூலறிவால் புதுப்புது தொழில்நுட்பத்தை நடைமுறைபடுத்தினாலும் இயற்கையின் தன்மையை அறிந்து அதற்கு எந்தவிதமான மாறுபாடு அல்லது தடங்கல் இல்லாமல் நம்முடைய தொழில்நுட்பத்தைப் புகுத்தி வளர்ச்சியடையச் செய்யவேண்டும். இயற்கைக்கு ஊறுவிளைவிக்கும் எந்த தொழில்நுட்பத்தை செயல்படுத்தினாலும் தற்பொழுது பல நன்மைகளை நாம் அடைவதுபோல் தோன்றினாலும் எதிர்காலத்தில் மனித இனமே ஏன் எந்த உயிரினமுமே உலகத்தில் வாழ்வதற்கற்ற சூழ்நிலை உருவாகிவிடும்.

When we develop any new technology for our comfort based on our educational knowledge, we should implement the same without disturbing our nature/environment. Though some of the new technologies seems to be beneficial at present, it may lead to environmental degradation in future and could be the cause for the extinction of life on earth. Our main goal is to safe guard all the living beings from extinction.

When human civilization started, so many technologies were developed all over the world to fulfil their basic needs. People started to exploit the natural resources for their convenience. Every development enhanced the quality of the human life but at the same time it degraded our environment and paved the way for depletion of natural resources.

Whenever a new technology is developed, it should be in a sustainable manner. "Sustainable development is a development that meets the needs of the present goal without compromising the ability of future generations to fulfill their own needs." Sustainable development can be simply described by the elasticity of the rubber, when rubber is stretched within the limit of elasticity, it can be reused again and again. If we stretch the rubber beyond the limit then it will be permanently deformed and it will lose its quality. Hence when we develop any technology, we must think about its impact on environment in future. Our development must be within the limit of regaining capacity and tolerable limits of the environment and natural resources.

COMPONENTS OF EARTH

Lithosphere:

The lithosphere is the outermost sphere of the Earth and it is important largely because it is the area that the land dwellers of biosphere inhabit and live upon. It is also the area of resource, almost all of our resources like coal, oil, gas and elements like iron, aluminium, calcium, copper, magnesium, etc which have widespread uses for the mankind.

Hydrosphere:

Hydrosphere covers all the area where water is present on the Earth surface. Water is necessary for sustaining life on Earth.

Atmosphere:

The atmosphere consists of a blanket of gases which surrounds Earth. It is held near the surface by Earth's gravitational attraction. Without atmosphere, there could be no life on Earth. The atmosphere maintains the climate on Earth. Earth's atmosphere is moderate and more affordable than any other planets.

Biosphere:

The biosphere is the layer of the Earth where life exists. All the microbes, plants, and animals can be found in the biosphere. The biosphere extends between the upper areas of the atmosphere up to 10 kilometers from the sea level where some birds and insects can be found and to the greater depths of the ocean at more than 6 kilometres.

All the three spheres are working together continuously for the benefit of the biosphere and all the spheres are inter linked with each other. When any sphere is disturbed or polluted, it will reflect in the other spheres. Normally all essential life supporting components are cycling among all the spheres continuously. These cyclic process are commonly called as Biogeochemical cycle.

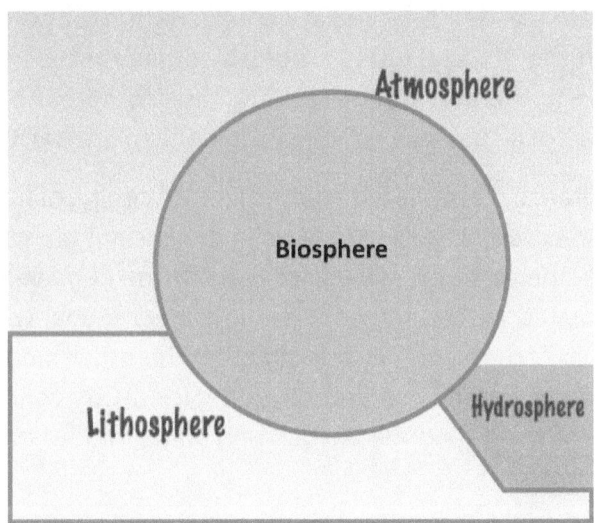

Biogeochemical Cycle:

Elements like carbon, hydrogen, nitrogen, oxygen, phosphorus, calcium and sulfur, molecules like water and silica are continuously moving from one sphere to all the spheres in cyclic manner. Through these cycles, essential nutrients are entering into the biosphere (bio organisms) from the lithosphere, atmosphere, and hydrosphere for the biological process and similarly move from the biosphere to other spheres.

The speed of the cycles and the concentration of chemicals in each sphere were maintained in a stable manner till the pre industrialization period. But with the advent of industrialization, due to human intervention, the speed, intensity, and concentration of chemicals in these cycles changed drastically. These changes affected the life of bio organisms and many of them went extinct. Here we will analyse some of the Biogeochemical cycle and how they are affecting the environment.

Carbon Cycle:

Carbon enters as carbondioxide from atmosphere to the biosphere through the process of photosynthesis and this carbon in CO_2 is converted into glucose and stored in the plant tissues. Then these carbon as glucose is transferred from plants to herbivores (plant eaters) and then carnivores (animal eaters) via food chain. Finally by respiration or the death of bio organisms (animal and plants), the carbon from the glucose gets aerobically or anaerobically degraded and released to the atmosphere as CO_2, thus completing the carbon cycle. By this cyclic process, all the bio organism are getting energy.

This natural carbon cycle is disturbed by the burning of huge quantity of fossil fuels, Crude oil, coal, lignite etc – the ancient bio carbon which was stored in the earth for millions of years. These huge venting of CO_2 gets accumulated in the atmosphere and trigger global warming. This disturbance in carbon cycle is also affecting the other biogeochemical cycles especially water cycle and oxygen cycles. CO_2 concentration is responsible for regulating the oxygen and water vapour concentration in the atmosphere.

Oxygen Cycle:

Oxygen enters from atmosphere to biosphere through respiration and released to the atmosphere during the photosynthesis. Whenever CO_2 concentration increases in the atmosphere, oxygen concentration decreases. Because every 44 grams of CO_2 produced during the burning of fossil fuels lead to depletion of 32 grams of oxygen from atmosphere.

$$C \quad + \quad O_2 \quad \longrightarrow \quad CO_2$$
$$\text{(12 grams)} \quad \text{(32 grams)} \quad . \quad \text{(44 grams)}$$

Likewise if we reduce the CO_2 concentration in atmosphere, automatically oxygen concentration will increase.

Water Cycle:

Water from the oceans and other bodies enter to the atmosphere by evaporation, again return back to the oceans by precipitation. In this cycle, direct rain on land and run off water through the rivers and streams are used by the biosphere. Water vapour concentration also will increase in the atmosphere due to the Increasing CO_2 levels. More CO_2 levels in the atmosphere increases the global warming and induce more evaporation thereby more water will be shifted from the water bodies to the atmosphere as water vapour. For example, due to global warming if the surface temperature of the earth increases to 100°C then all the water in the ocean will get accumulated in the atmosphere and the ocean will be dried out. Like this, high CO_2 level in the atmosphere affects entire natural Biogeochemical cycles. If we normalise the level of carbon concentration in the atmosphere, then all other natural process will be normalised. Likewise other biogeochemicals such as sulfur and phosphorus cycles are also affected by this carbon intervention, since huge amount of sulfur and phosphorus are entering into atmosphere due to burning of fossil fuels.

Unsustainable Development:

Producing thermal power using fossil fuels without evaluating the consequence is considered as unsustainable development. This type of development will increase global warming, sea level rise, acid rain and oxygen depletion. All these environmental degradation activities will wipe out the bio organisms from existence.

Unsustainable Development

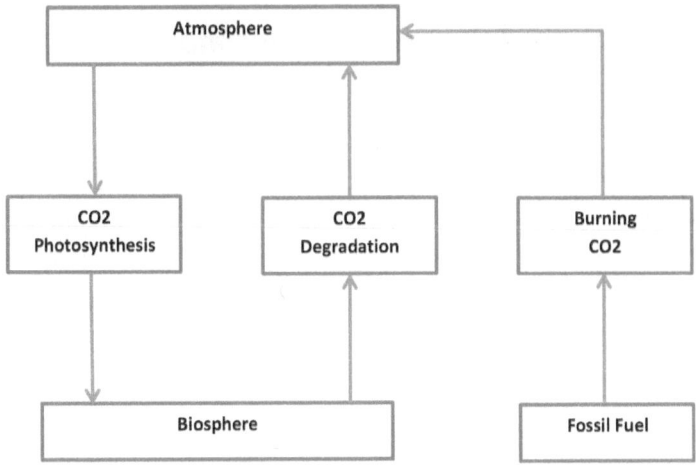

Continuous accumulation of Co_2 in to the atmosphere

Sustainable Development:

Producing energy by using fossil fuels and avoiding the CO_2 emissions into the atmosphere is considered to be sustainable development. It can be possible by adapting the Carbon capture and sequestration technology in thermal power plants.

Sustainable Development

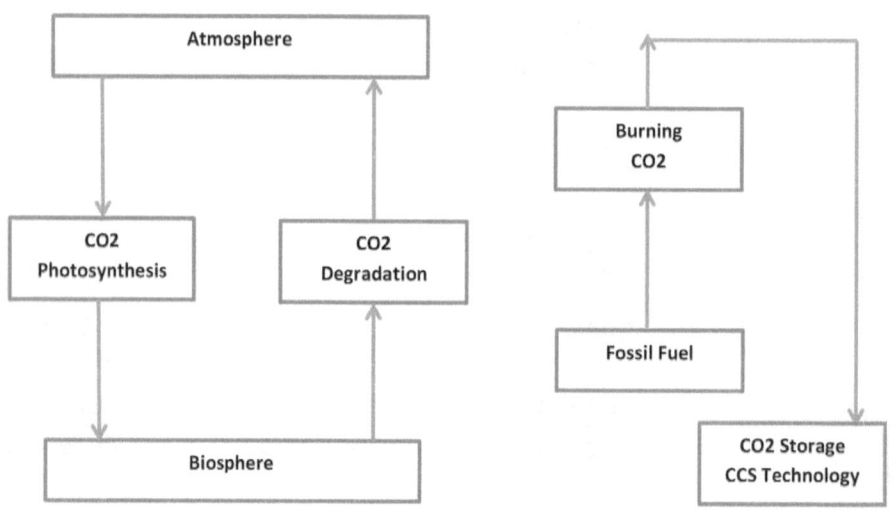

No accumulation of co_2 in the atmosphere

Carbon Capture and Sequestration Technology:

Carbon capture and sequestration is the process of capturing CO_2 from the exhaust chimney of thermal power plants, cement industries and other CO_2 emitting industries. The captured CO_2 can be either kept in deep underground for long term storage or used for specific purposes. Carbon capture and sequestration technology can be subdivided mainly into three categories namely physical, chemical and biological methods.

Physical or Geological Method:

Captured CO_2 will be liquefied and then stored in void areas already created inside the ground during the process of mining the fossil fuels. CO_2 can be easily liquefied as its critical temperature, and critical pressure can be easily achieved. Critical temperature of carbondioxide is 31°C, critical pressure 72.9 bar. The liquefied CO_2 can be injected through pipe line into deep ground with concrete housings. This liquefied CO_2 can also be stored in the bottom of the ocean floor at the depth beyond 6000 meter from the sea level, where it will stay comfortably at 4°C and 600 bar. By this method, a larger volume of gaseous CO_2 can be stored in very limited space in liquid form.

Biological Method:

There are so many technologies available for capturing CO_2 by biological method. But foremost of them is plantation of fast growing plants in the vicinity of the thermal power plants. The plants will take more CO_2 for their photosynthesis processes. Likewise we can also introduce plantation technology in nearby coastal regions.

(The world record for the fast growing plant blongs to certain species of the 45 varieties of bomboo, which have been found to grow at up to 91 cm per day.)

Chemical Method:

Captured CO_2 by the CCS technology can be utilized for manufacturing so many organic and inorganic compounds. Organic compounds like urea, formic acid, carboxylic acids, methanol and variety of organic products can be produced by utilizing CO_2. Likewise so many mineral carbonates can also be produced with CO_2 and stored over a long period as mineral carbonates have more stability.

Example

$$Mg_2SiO_4 + 2CO_2 \longrightarrow MgCO_3 + SiO_2.$$

The captured CO_2 can also be used directly in fire extinguishers, refrigerators and making carbonate beverages in food industries

The cost of adaptation and implementation of the carbon capture and sequestration technology (CCS) in thermal power plants is very much affordable and cost effective compared to the cost of revival of the degraded environment.

We installed power plants and other heavy industries some 100 years ago based on our requirements and designed the technology based on our educational knowledge. However we were not concerned about the wellness of our environment and nature at that time of introduction, as mentioned in this kural. If we had foreseen the subsequent effects of these power plants on environment, we would have found the solution much earlier for removing the excess CO_2 accumulation into the atmosphere and would have avoided all environment degradation.

Now worldwide, people are afraid of "CORONA" virus. It creates big threat and induces much fear in people's minds. Around the globe there is a recession in economy. Knowingly or unknowingly this disaster is caused by human.

> எதிரதாக் காக்கும் அறிவினார்க் கில்லை
> அதிர வருவதோர் நோய்.

> No terrifying disaster will happen to a wise person
> who foresee and safeguard against the coming evils.

வரப்போவதை முன்பே அறிந்து காத்துக் கொள்ளவல்ல அறிவுடையவர்க்கு, அவர் நடுங்கும் படியாக வரக்கூடிய துன்பம் ஒன்றுமில்லை.

"Sustainability creates and maintains the conditions under which Humans and nature can exist in productive harmony"

2 BIODIVERSITY

பகுத்துண்டு பல்லுயிர் ஓம்புதல் நூலோர்
தொகுத்தவற்றுள் எல்லாம் தலை.

Sharing food and taking care of lives, is the foremost amongst all virtues compiled in all scriptures.

நம் வாழ்க்கையில் செய்யும் அனைத்து தர்மங்களிலும், நியாயங்களிலும் நாம் உழைத்து சம்பாதித்த உணவை மற்றவர்களுடன் பகிர்ந்து உண்பதே சிறந்த அறம் ஆகும். இங்கு திருவள்ளுவர் அனைத்து மனிதர்களுடன் பகிர்ந்து உண்பதே உன்னதமான வாழ்க்கை என்று மட்டும் கூறவில்லை, "பல்லுயிர் ஓம்புதல்" என்றும் குறிப்பிடுகிறார். அதாவது அனைத்து உயிர்களுக்கும் உணவு அளிக்க வேண்டும், அனைத்து உயிர்களையும் போற்றிப் பாதுகாக்க வேண்டும் என்று கூறியிருக்கிறார். இந்தக் குறளின் மூலம் அவர் "பல்லுயிர்பெருக்கத்தின்" அவசியத்தை வலியுறுத்திக் கூறியுள்ளார்.

The food whatever we earned and collected has to be distributed to other organisms equally. leading the life in such a way will be the most meaningful and honest. Humans completely depend on other organisms for their food, dress, shelter and medicine. Not only humans, every organism in the universe depends on the other organisms for their survival. Each and every organisms are interlinked with the food chain. An organism may be a prey for another organism and predator for some other one. If one organism is extinct (dead) in the biosphere due to food scarcity or some other environmental degradation, then the entire biosphere will be affected

Here our great poet Thiruvalluvar described about the importance of biodiversity and its importance in the environment.

Biodiversity:

"No man is an island" This statement is also true for any organism in the ecosystem. No organism can exist in isolation. Individual organisms lives together in an ecosystem and depend on each other. In fact they have various levels of interaction with each other and many of these interactions are critical for their survival.

Biodiversity is the shortened form of two words "biological" and "diversity". It refers to all the variety of life that can be found on Earth (plants, animals, fungi and micro-organisms) as well as to the communities that they form and the habitats like terrestrial (land), marine and other aquatic ecosystem. Approximately there are 8.7 million species in earth out of which 6.5 million species are in the land and remaining 2.2 million species are in the ocean. The area which is having a variety of plant and animal species will be considered as rich in biodiversity. Biodiversity richness is a very significant indicator for the classification of an area into various ecosystems among all other parameters.

Cluster of plants and animals including microorganisms lives together in a particular area like forest ecosystem, grassland ecosystem, pond ecosystem and desert ecosystem, etc.

In the forest ecosystem, trees are producing food to the entire animals living in the forest with the help of Sunlight, CO_2, water and other nutrients. The food produced by the trees will be transferred from one organism to the other organisms in the food chain. Low level animals like rabbits and deer eat plants, higher level animals like fox and wolf eat rabbits and deer. Ultimately the higher level animals like lion and tiger eat wolf and Fox. Finally, on death of all organisms, their bodies are degraded by the microorganism into CO_2 and water. In such a way the forest ecosystem exists continuously without degradation.

In the food chain, if any one of the organism goes extinct then the entire forest will be affected to a great extent. For example, if lion and tiger population are reduced by environmental degradation, then the rabbit and deer population will grow drastically and they will spoil the entire forest by over grazing. In the same way if rabbits and deers are extinct by over hunting or by any other reason then the lion and tiger will not get food in the forest and they will enter into the human residential areas

Small creatures like insects, bees, moths and butterflies, habituating the forest, consumes nectar from the flowers. These small insects are eaten by the birds, then the birds are eaten by the vulture and eagle. If all these mini creatures goes extinct then the forest will be no more due to lack of pollination and lack of seed dispersal activities

Biodiversity is essential for the welfare of the present and future generation. Sustainable biodiversity is sustainable development. Each and every creature in the globe has its own importance and service to the society by their routine activities. If one creature disappears at low level or higher level, its effect will be reflected throughout the entire world.

For example, in Mauritius island, the population of a tree named "dambalacoke" got drastically reduced and it was included in the extinct organisms list. At present the number of dambalacoke trees are only thirteen. Scientists took various measures to grow and increase their numbers without much success. The seed of the trees could not get sprouted by any means.

It was much later that scientists found out the real reason. There existed a bird called "tutu" which were largely hunted for food by the people living there. Due to excessive hunting, the bird went extinct. The decreasing population of the dambalacoke trees are directly linked to the extinction of "tutu" birds. There was a distinct relationship between the trees and the "tutu" birds. The birds have to eat the fruit of the tree and then the seed which gets excreted alone has the ability to grow as a Tree. Thus the extinction of the bird caused the extinction of the dambalacoke trees.

Similarly in Africa, a tree named "ombulocarbum elattam" is related with the elephants. The seed of ombulocarbum elattam grows only after passing through the digestive system of the elephants. The extinction of elephants or diversion of the path of the elephants because of human activities, will lead to the extinction of these trees.

Likewise humans will also be in the extinction list if the following organisms go extinct.

Ants:

Ants aid in decomposition and turn up more soil than earthworm. When ants dig tunnels they aerate the soil and recycle nutrients. This activity is ecologically crucial in maintaining healthy soil for plant growth. Ants help to reduce the use of chemical fertilizers and the need for irrigation. A study conducted in 2011 concluded that ants help to increase wheat crop yield by 36%. Ants also are effective pollinators like bees and butterflies.

Bees:

Bees are the super stars of the pollination. They are probably the finest creature than any other creature that is responsible for pollinating flowers. Approximately one third of our food supply depend on the cross pollination and that too by bees. If bees go extinct then the food production could be very critical. we could not even get a single fruit and vegetable. In addition to pollination, bees collect honey which has much medicinal value.

Bats:

Bats are exceptionally important in our ecosystem. Bats are insectivorous, they perform insect control service by consuming millions of pests each year. Farmers are grateful to bats because they save millions of money each year by reducing the need of chemical pesticides. In many places of the world, mosquitoes are vectors of deadly diseases such as malaria and dengue fever. A single brown bat can eat upto 1000 mosquitoes in one hour.

Bats are insectivorous as well as fruit eaters, so they disperse seeds to other areas by their excreta. Bats travel for long distance and help plants to grow in various locations. Due to their effective seed dispersal service, bats are called as farmers of the forest.

Bats also help in pollination of flowering plants. Nectar eating bats are crucial pollinators for over 500 plant species. Most of the flowering plants do not have the ability to produce seeds without being pollinated.

Birds:

Birds perform a broad variety of ecological role including afforestation, insect control, nutrient recycling, plant pollination, seed dispersal and as the bio indicators of ecosystem. Some ground-dwelling birds even help aerate and turn up soil with their claws. Ants and bees might be the masters of their own trades but birds are masters of all the trades.

காக்கை கரவா கரைந்துண்ணும் ஆக்கமும்
அன்னநீ ரார்க்கே உள

காக்கை தனக்குக் கிடைத்த உணவை, தனியே உண்ணாமல் மற்றவர்களுக்கும் பிரித்துக் கொடுத்துவிட்டு உண்ணும் உயர்ந்த பண்புடையது. அதுபோல நம் உணவை பிற உயிர்களுக்கும் மற்றவர்களுக்கும் பகிர்ந்துண்டு மற்ற உயிர்களையும் பேணி பாதுகாத்து வாழும் வாழ்வானது சிறந்த வாழ்க்கை முறையாகும். அப்படி உயர்ந்த குணம் இருந்தால் அவர்களிடம் செல்வம் மேலும் மேலும் வந்து சேரும்.

காக்கை குருவி எங்கள் ஜாதி பெரும்
காடும் மலையும் எங்கள் கூட்டம்.

மகாகவி பாரதியார் கூட "காக்கை குருவி எங்கள் ஜாதி" என்று கூறியிருக்கிறார் இதன் மூலம் அனைத்து உயிர்களையும் பேணி பாதுகாக்கவேண்டும் என்று வலியுறுத்துகிறார்.

Conclusion:

Our way of life should safe guard all the living creatures around us. We should be very cautious when we introduce a new technology or new chemicals such as fertilizers and pesticides. It should not give negative impact to any living organism and should be eco-friendly in nature.

Even a single organism should not be affected by the development. If we destroy other organisms for our own comfort, then finally we (human) also will be washed away from the earth.

> "Protecting biodiversity is just as important and critical to the survival of mankind as stabilizing the climate. Species protection and climate are interdependent."
>
> — Klaus Topfer

3 ADVERSE EFFECTS OF OZONE IN THE TROPOSPHERE

தலையின் இழிந்த மயிரனையர் மாந்தர்
இழிந்த நிலையின் கடை

When people slip from their dignified position
they are treated lowly as a hair fallen from head.

நம் தலையிலுள்ள முடியானது நமக்கு மிகுந்த அழகை கொடுக்கின்றது. அதே முடி நம் தலையில் இருந்து கீழே விழுந்து விட்டால் அதைப்பற்றி யாருமே கவலைப்பட மாட்டார்கள். அந்த முடி நாம் உண்ணும் உணவிலோ அல்லது குடிக்கும் நீரிலோ கலந்துவிட்டால் அது நமக்கு மிகுந்த அருவருப்பைத் தரும். மற்றும், உணவின் வழியாக முடி வயிற்றில் சென்றுவிட்டால், அது பெரும் துன்பத்தைத் தரும். அது போலவே ஓசோன் படலமானது வாயுமண்டலத்தில் மேலே இருக்கும் போது மனித குலத்திற்கு மிகுந்த நன்மையை தருகிறது. அதே ஓசோன் கடல் மட்டத்தில் இருக்கும்போது சுற்றுப்புற சூழலுக்கும் மனித குலத்திற்கும் மிகுந்த துன்பத்தைக் கொடுக்கின்றது.

The hair when it is in the head will be treated preciously by people as it enhances beauty and handsome look. Apart from good appearance it also plays a vital role in maintaining ones good health. The hair which present in the head serves primarily as a source of heat insulation as well as protection from ultra-violet radiation exposure. The hair in the human body regulates and maintains internal body temperature at required level. When the body is too cold, the arrector muscles found attached to hair gets erect. These hairs then form a heat-trapping layer above the epidermis. The opposite actions occur when the body is too warm, the arrector muscles make the hair lie flat on the skin which allows heat to leave. This process is formally called piloerection. So people takes more care when it is on their head but at the same time if it is fallen from the head, the same hair gives ugly appearance and people get scared as it may enters in our body through food and damages the digestive system. They ll get rid of it as fast as they can.

In the same way, ozone, when it is in the stratosphere region (top) acts as an umbrella and protect us from the harmful ultraviolet rays. When the same Ozone descends from its position to Troposphere which is near to the earth surface (Sea level), it will give adverse effect to the mankind as well as to other living organisms including plants.

What is ozone?

Ozone is tri atomic oxygen molecule. It is formed in the upper atmosphere by the interaction of UV rays with the oxygen molecule.

Ozone is a gas that occurs both in the Earth's upper atmosphere and at ground level. Ozone can be "good" or "bad" for our health and the environment depending on its location in the atmosphere

Ozone formation (Ozone Oxygen cycle):

An oxygen molecule in the stratosphere region undergoes photo dissociation by high frequency UV rays. This creates two oxygen free radicals.

$$O_2 \text{ (g)} \xrightarrow{\text{(UV rays)}} 2O \cdot \text{(g)}$$

The oxygen free radical combine with another oxygen molecule and then form ozone gas

$$O_2 \text{ (g)} + O \cdot \text{(g)} \xrightarrow{\text{(UV rays)}} O_3 \text{ (g)-}$$

The ozone molecule absorbs UV radiation, dissociates into oxygen and oxygen free radicals.

$$O_3 \text{ (g)} \xrightarrow{\text{(UV rays)}} O \text{ (g)} + O_2 \text{ (g)}$$

Any remaining oxygen free radicals reacting with another ozone forming oxygen gas, and thus the cycle continues further.

$$O_3 \text{ (g)} + O \cdot \text{(g)} \rightarrow 2O_2 \text{ (g)}$$

In this way ozone concentration is maintained continuously. At any time the ozone concentration is constant in this ozone oxygen cycle.

Benefits of Ozone in Stratosphere Region:

High-Altitude "Good" Ozone

In higher altitudes the ozone layer protects us as an umbrella from the powerful ultraviolet rays. The ozone layer absorbs 97 to 99 percent of the Sun's medium-frequency ultraviolet rays (from about 200 nm to 315 nm wavelength), which otherwise would potentially damage life of living things near the surface of the earth.

Radiation:

Radiation is the emission of energy from any source. There are many types of radiation ranging from very high-energy (high-frequency) radiation – like x-rays and gamma rays – to very low-energy (low-frequency) radiation – like radio waves. UV rays are in the middle of this spectrum. They have more energy than visible rays but not as much as x-rays.

Sun emits radiation which are UV rays, visible rays and Infra-red rays, out of which the UV rays create more ill effects when it reach the earth's surface.

There are different types of UV rays based on the energy they have

Type of UV Rays	Wave Length	Energy
UV-A	315–400 nm	less energy
UV-B	280–315 nm	more energy than UV A
UV-C	100–280 nm	high energy UV rays

Higher energy UV rays are considered as ionisation radiation. This means they have enough energy to remove an electron from an atom or molecule. Ionising radiation can damage the DNA (genes) in cells which in turn may lead to cancer.

(Fortunately even the highest-energy UV rays don't have enough energy to penetrate deeply into the body beneath the skin, hence their target is limited to the skin. If it has more penetrating power, it will cause cancer to entire internal organs of our body)

The ozone layer contains less than 10 parts per million of ozone, while the average ozone concentration in Earth's atmosphere as a whole is about 0.3 parts per million. The ozone layer is mainly found in the lower portion of the stratosphere, from approximately 15 to 35 kilometers above the sea level. Its thickness varies seasonally and geographically.

If there was no ozone in the stratosphere region, then increased penetration of UVC and UVB can lead to skin cancer and cause considerable damage to all the plants and animals.

Short-wavelength UVC is the most damaging type of UV radiation. However, it is completely filtered by the ozone layer in the stratosphere region and does not reach the earth's surface.

UVC is the high potential dangerous radiation to the environment. No UV-C radiation reaches to earth surface from the sun. However some people are exposed to the UVC radiation when they work with welding torches or mercury lamps.

The strength of the UV rays reaching the ground depends on number of factors as mentioned below

UV rays are strongest between 10 a.m. and 4 p.m. in a day and are stronger during spring and summer season. UV exposure goes down as you move farther from the equator. More UV rays reach the ground at higher elevations. The effect of clouds can vary, but what's important to know is that UV rays can get through to the ground, even on a cloudy day. UV rays can bounce off surfaces like water, sand, snow or even grass leading to an increase in UV exposure.

HARMFUL EFFECTS OF OZONE WHEN IT IS NEAR THE EARTH'S SURFACE

(Troposphere):

In the troposphere region, the ozone acts as a green house gas and contributes to global warming. Its contribution towards global warming as greenhouse gas is around 3 to 7%. It is one among the three major green house gases next to methane.

Greenhouse Gases	Contribution towards Global Warming
Water vapour	36 to 72%
Carbon dioxide	24 to 26%
Methane	4 to 9%
Ozone	3 to 7%

Health Hazards:

How Does Ozone Affect Human Health and the Environment?

Breathing ozone can trigger a variety of health problems including chest pain, coughing, and throat irritation. It can worsen bronchitis, emphysema, and asthma. Ozone also can reduce lung function and inflame the linings of the lungs. Repeated exposure may permanently scar lung tissue.

Healthy people also experience difficulty in breathing when exposed to ozone pollution. Because ozone forms in hot weather, anyone who spends time outdoors during summer may be affected, particularly children, older people, outdoor workers, etc.

A statistical study of 95 large urban communities in the United States found significant association between ozone levels and premature death. Tropospheric Ozone causes approximately 22,000 premature deaths per year in 25 countries in the European Union. (WHO, 2008)

Effects on Plants:

Ground-level or "bad" ozone also damages vegetation and ecosystems. It leads to reduced yield in agricultural crop and commercial forests.

Effects of Ozone on Material:

Rubber, textile dyes, fibers, and certain paints may be weakened or damaged by exposure to ozone. Some elastic materials can become brittle and crack, while paints and fabric dyes may fade more quickly.

Formation of Photochemical Smog:

Ozone also leads to the formation of smog or haze causing additional problems such as a decrease in visibility as well as damage to plants and ecosystems. High UV penetration can become the catalyst for the formation of photochemical smog which occurs in the Troposphere.

Conclusion:

The ozone when it is in the top of the atmosphere, filters UV radiation and prevents entire bio organisms in the earth. If the ozone gets depleted in the stratosphere region then the biosphere (Entire animal and plants) suffers considerably. But if the ozone comes to ground level, it causes complete devastation to an extent that one could not even imagine.

"Don't expect same response from anyone always, they may be on vacation"

BENEFICIAL EFFECTS OF GLOBAL WARMING

கேட்டினும் உண்டோர் உறுதி கிளைஞரை
நீட்டி அளப்பதோர் கோல்.

There is a benefit in grief, it is a tool
To test and gauge a friend.

தீமை வந்தால் அதிலும் ஒரு நன்மை உண்டு. அந்தத் தீமைதான் நண்பர்கள் எப்படிப்பட்டவர்கள் என்று அளந்து காட்டும் கருவியாகிறது. தீமை ஒருவனுக்கு வந்தால் அவன் மிகுந்த துன்பம் அடைய நேரிடும். ஆனால் அந்த துன்பம்கூட ஒருவிதத்தில் நமக்கு நல்ல நண்பன் யார் கெட்ட நண்பன் யார் என்று அறிந்துகொள்ள உதவும். அதுபோலவே உலகவெப்பமயமாதல் என்பது பலப்பல அச்சுறுத்தல்களையும் பல சுகாதாரச் சீர்கேடுகளையும் விளைவித்துக் கொண்டிருக்கிறதெனினும், இந்த உலகவெப்பமயமாதலிலும் சில நன்மைகள் இருக்கின்றன.

When we suffer from any problem, we also gain something out of it. That problem alone acts as an instrument to measure the quality of our friends.

Similarly Global warming also creates so many adverse effects on the environment such as sea level rise, destruction of biodiversity, heat waves, food scarcity etc. However we get some benefits also through global warming.

Let us see the positive effects of global warming.

Agriculture:

Global warming in Greenland brings longer and warmer summer. This can pave the way for new possibilities for agriculture in that region. Improved agriculture and increased growing season in some high latitude regions like Greenland, increases productivity of sour orange trees. In southern Greenland, the summer period is increased by three weeks compared to a decade ago by global warming . This longer and warmer growing seasons benefits sheep farmers in south Greenland. They could grow more grass and for a longer period, which is important for their cattle production. It gives more possibilities for producing new types of vegetation in that region. For agricultural production, the improved conditions are resulting in a larger plant yield and new opportunities for commercial production of potatoes and vegetables.

Fishing:

As ice melts, the area of water surface also increases and hence the fishing industries in that particular area expands. Greenland economy is getting improved due to the enhanced fishing activities.

Health:

More people die during cold winter months than in the warm summer months. In European countries, the average number of deaths in winter (December – March) is 10% to 30% more than the summer. The reason for these excess death in winter is directly related to low ambient temperatures. In that way, global warming decreases the rate of deaths in winter.

Shortest Sea Routes:

An ice-free Northwest Passage, provides a shipping shortcut between the Pacific and Atlantic Oceans. The depletion of the arctic ice is already visible. Various projections indicate that the ice sheet will greatly diminish in the future, and might even disappear by the end of this century. On average, the ice sheet has diminished by 4.3% every ten years. By 2045 to 2060, the decline of Arctic sea ice under moderate warming could allow even ordinary cargo ships to travel directly over the North Pole.

As Arctic routes become more direct, voyage times could fall to less than three weeks in some cases, making Arctic shipping potentially more attractive than the southern routes in coming decades. These are all the positive effects of global warming. However global warming creates so many adverse effects on climate as well as in biosphere., our present goal has to be the protection of the polar ice caps and not finding the shortest ship routes.

"எந்தநன்மைக்குள்ளும் சில தீமையுண்டு
எந்தத்தீமைக்குள்ளும் சில நன்மையுண்டு"

"In each good thing there are some bad effects And
In each bad thing there are some good effects too"

5 | SELFLESSNESS OF EARTH

இரத்தலின் இன்னாது மன்ற நிரப்பிய
தாமே தமியர் உணல்.

Eating alone to increase one's hoarded wealth,
Is more awful than begging.

பிறருக்கு ஈவதால் குறையக்கூடும் என்றெண்ணி சேர்த்து வைத்துள்ளதை
யாருக்கும் கொடுக்காமல் தானே தனியாக உண்டு உயிர் வாழ்வதென்பது
இரந்து வாழ்தலைவிட, அதாவது யாசகம் எடுத்து உயிர் வாழ்தலைவிடக்
கொடுமையானது.

Some of us believe that our wealth will diminish when it is given to the needy people and so we consume everything by ourselves without sharing. This poor attitude is considered as horrible way of life than begging. This will destroy all our wealth very shortly and finally leads us to a pathetic condition.

This can be compared with the energy sharing concept of earth and solar system. Earth is receiving plenty of energy from the sun. If earth keeps all the energy within it and consume everything, it might have been burnt into ashes some million years ago like a selfish greedy man.

Let us calculate the rise in temperature of the earth, if it keeps all the energy that falls on its surface.

Mass of earth = 5.9722×10^{24} kgs
Specific heat of water = 4180 joule/kg/k

(Around two third of the earth surface is covered with water, more over specific heat of the land mass is less than the water, hence specific heat of the water alone is taken for calculation for entire earth temperature rise)

Energy Receives by the Earth:

Earth surface	=	$5.1 \times 10^{14} \, m^2$

Solar radiation per second per m^2
on the earth surface = 342 joule

Total heat received by the earth
per second

$$= 5.1 \times 10^{14} \times 342 \text{ joules}$$

$$= 1744.2 \times 10^{14} \text{ joule/second}$$

Total heat received by the earth
per hour

$$= 1744.2 \times 10^{14} \times 3600 \text{ joule}$$

Total heat received by the earth
per year

$$= 1744.2 \times 10^{14} \times 3600 \times 24 \times 365$$

$$= 5.5 \times 10^{24} \text{ joule/year}$$

Quantity of heat (Q) = m × s × t
(m = mass, s = specific heat and t = temperature.)

t = q/m × s

$$\text{Temperature rise per year} = \frac{5.5 \times 10^{24}}{5.9723 \times 10^{24} \times 4180}$$

$$= 0.00022° \, c$$

$$\text{For raising } 100° \, c = \frac{1}{0.00022} \times 100$$

$$= 454545 \text{ years}$$

Earth was formed around 4.54 billion (45400000000) years ago. As per the above calculation, if the earth keeps all the heat within it and does not spend for other natural processes such as hydrological cycle, air cycle and photosynthesis then it would have been vanished by solar radiation some billion years ago.

வைத்தான்வாய் சான்ற பெரும்பொருள் அஃதுண்ணான்
செத்தான் செயக்கிடந்தது இல்.

Wealth all over the house stashed, but not used
Makes one insignificant as if he is deceased.

தன் வீடு நிறையப் பெரும்பொருள் சேர்த்து வைத்திருந்தும்,
கஞ்சத்தனத்தால் அதை அனுபவிக்காதவனுக்கு அப்பொருளால்
பயனில்லை. ஆதலால் அவன் இருந்தாலும் இறந்தவனுக்குச் சமம்.
பூமி, சூரியனிடமிருந்து பெற்ற அனைத்து ஆற்றலையும், எல்லா
உயிரினங்களுக்கும் பயன்படக்கூடியவகையில் மழையாகவும்,
காற்றாகவும், உணவாகவும் மாற்றிக்கொடுத்து வாழவைத்துக்
கொண்டிருக்கிறது. பூமி தான்பெற்ற அனைத்து ஆற்றலையும் தானே
வைத்துக்கொண்டிருந்தால், பூமி என்றோ அழிந்திருக்கும் அல்லவா?

Even if someone possess abundant quantity of wealth but have not
utilized the same because of their stingy behavior, then there is no use
for him by that wealth. That person will be treated as a dead person
though he is alive. Earth exists by giving all the energies that are received
from the sun to all the living things as rain, wind, food, etc.

> *"No one has ever became poor by giving"*
>
> — *Anne Frank*

6 | IMPORTANCE OF CONSERVATION TILLAGE

ஏரினும் நன்றால் எருவிடுதல் கட்டபின்
நீரினும் நன்றதன் காப்பு.

Worthier than ploughing is to fertilize the field and
After weeding, Worthier than watering is to secure the yeild.

நிலத்தை விவசாயத்துக்கு ஏற்றவாறு பக்குவப்படுத்த வேண்டுமெனில் ஏர் மிகவும் அவசியம். வெறுமனே உழுதால் மட்டும் போதாது தானே. நிலத்துக்குச் சரியான எருவிடவில்லை என்றால் பயிர் நன்றாகவளராது தானே. எனவேதான் ஏரைவிட எரு முக்கியம் என்று குறிப்பிட்டுள்ளார் வள்ளுவர். மற்றும் எரு இட்டபின் நீர்ப் பாய்ச்சுதலும் களையெடுத்தலும் பயிர் நன்றாக வளர உதவும். இறுதியாக, இவையெல்லாவற்றையும்விட வேலி போட்டு பாதுகாப்பது மிகமிக அவசியமென்று விவசாயத்தைப்பற்றி தெளிவாக விவரித்துள்ளார்.

ஏரை அடிக்கடி உழுவதாலும் மேலும் இயந்திரத்தைக்கொண்டு மிகவும் ஆழமாக உழுவதாலும் நிலத்தில் உள்ள சில சத்துக்கள் ஆவியாகி சென்றுவிடும் என்று உணர்ந்து, ஏரைவிட எருவிட்டு நிலத்தை நன்றாகப் பாதுகாக்கவேண்டும் என்று உழவைப்பற்றி குறிப்பிட்டுள்ளார்.

The main part of agriculture is tillage. Before seeding, tillage has to be done. Tillage makes the soil aerated and soft which is suitable for cultivation. But it should not be done beyond the necessary levels, because over tillage will dislodge the nutrients from the soil texture and send them to atmosphere.

Advantages of Tillage:

Tillage loosen and aerates the top layer of the soil, which facilitates in planting the crop smoothly. Tillage helps to mix the harvest residue, organic matter, humus and nutrients and evenly distribute in the soil. It also destroys the weeds mechanically. These are the prime benefits of tillage.

Disadvantages of Tillage:

Due to tillage, soil loses nutrients such as nitrates, sulphates and phosphate. Tillage enhances more exposure of the soil with sunlight and air. Nutrients lose the cohesion with soil and moves to the atmosphere. Tillage also reduces the microorganisms such as microbes, earthworms, ants, etc which are vital for crop yield.

Conventional Tillage:

Conventional tillage is the traditional method of farming in which soil is prepared for planting by completely inverting it with a tractor pulled plough, followed by subsequent additional tillage to smooth the soil surface for crop cultivation. Conventional tillage increases the CO_2 concentration in the atmosphere. While ploughing or tilling more soil is getting exposed in the sunlight which drives out more CO_2, NOX and other greenhouse gases into the atmosphere. The higher aeration in tilled soil increases the possibility of aerobic degradation in the soil and thus an increased potential for greenhouse gas emissions. Significantly higher N_2O emission has been reported from ploughed site than the non tilled sites .

This higher emission of N_2O from the soil will deplete the nitrate salt, but nitrate is very essential manure for the plant growth.

Conservation Tillage:

Conservation tillage, also called as Minimum tillage is a soil conservation system with the goal of minimum soil manipulation necessary for successful crop production. It prevents the emission of essential nutrients available in soil such as soil carbon, nitrates, sulphates, phosphates and other important components.

Biological activities are vital to the plants productivity and better yields. Hence the tillage must not affect the activities of microbes such as earth worms, termites and many other organisms in the soil.

In conservation tillage, there ll be limited tilling of the soil in order to provide better soil bed for saplings and aid their growth in initial period. Therefore, to achieve sustainable yield with minimal impact on the soil and atmosphere, conservation tillage would be paramount.

The best management practices usually entail the least amount of tillage necessary to grow the desired crop. This not only involves substantial saving in energy costs, but also conserve soil resources such as organic matter, humus, microbes, etc.

There is still some confusion whether the tillage is beneficial or not. Tillage can not be avoided completely in cultivation but it should be done as conservation tillage. More over frequent tillage leads to spending more energy and thereby increasing the cost of production.

Thiruvalluar also specified that conservation tillage, means minimum tillage that is required only for the sake of initial growth of seed and for primary arrangements for saplings to grow. Hence for the conservation of soil structure and microbes, too much of frequent tillage and deep tillage, better be avoided.

> "Knowledge, like money and manure serves us best when spread evenly"
>
> – Stuart Aken

7 OXYGEN DEPLETION

இரந்தும் உயிர்வாழ்தல் வேண்டின் பரந்து
கெடுக உலகையியற்றி யான்

Having to beg for survival, if anyone is so fated,
The maker of this world be damned.

வள்ளுவப் பெருந்தகை இந்தக் குறளின் மூலம் ஒருவன் யாசகம் செய்துதான் உயிர் வாழ வேண்டும் எனில் இந்த உலகை இயற்றியவன் கெட்டு ஒழிய வேண்டும் என்று ஆண்டவனுக்கே சாபமிடுகிறார். மனிதன் உயிர் வாழ்வதற்கு உண்ண உணவு, உடுக்க உடை, இருக்க வீடு இவை மூன்றும் மிகவும் இன்றியமையாததாகும். இதில் தலையாயது உணவு. உணவை ஒருவன் யாசகம் பெற்று தான் வாழ வேண்டுமெனில் இந்த உலகம் தேவையில்லை அதை உருவாக்கியவன் கூட தேவையில்லை என்று அளவுகடந்த கோபத்தின் உச்சிக்கே சென்று எச்சரிக்கைச் செய்துவிட்டு சென்றிருக்கின்றார். ஆனால் இன்று உணவை விட முக்கியமான சுவாசிக்கும் காற்று, குடிக்கும் நீர் போன்றவைகளே மனிதனுக்குத் தங்குதடையின்றி கிடைக்கக்கூடிய வாய்ப்பில்லை. அதாவது குடிக்கும் நீரையே காசு கொடுத்து வாங்கிக் கொண்டிருக்கிறோம். இன்னும் சிறிது காலம் போனால் சுவாசிக்கும் காற்றையேகூட காசு கொடுத்து வாங்கி சுவாசிக்கும் நிலைமை ஏற்பட்டுவிடும். இன்று திருவள்ளுவர் இருந்திருந்தால் உலகத்தைப் படைத்தவன் மட்டும் அல்லாமல் உலகத்தையே சாபமிட்டு அழித்திருப்பார்.

இதையேதான் மகாகவி பாரதியார் கூட, தனி ஒருவனுக்கு உணவில்லை எனில் ஜகத்தினை அழித்திடுவோம் என்று அவரும் வள்ளுவப் பெருந்தகையின் கூற்றை ஆமோதித்து அவருடைய கோபத்தையும், உலக சமுதாயத்தின் மீது காட்டி விட்டுச் சென்றிருக்கிறார். இவர்கள் இரண்டு பேர்களின் கருத்துக்கள் என்னவென்று கூர்ந்து கவனித்தால் இந்தப் பூமியில் பிறந்த ஒவ்வொருவனுக்கும் பூமியிலுள்ள இயற்கை வளங்கள் அனைத்தும் சொந்தம். அவனுக்கு எல்லா அத்தியாவசியப் பொருட்களும் தங்கு தடையின்றி கிடைக்க வேண்டும். அப்படி கிடைக்கவில்லை என்றால், அது அவனின் தவறு கிடையாது.

சமுதாயத்தின் தவறு. எனவே சமுதாயம் அதனுடைய தவறைத் திருத்திக் கொண்டு எல்லோருக்கும் எல்லாம் கிடைக்க வேண்டும் என்ற அடிப்படை உரிமையை நிலை நாட்டவே இரண்டு புலவர்களும் மேற்கண்டவாறு தங்களது கோபத்தை வெளிப்படுத்தியுள்ளனர்.

Thiruvalluvar says in this kural that begging for food to sustain one's own life is the most painful thing in the world. In that pathetic condition, he curses the God for this situation and states that there is no need for the presence of God and the God may destroy himself.

In the same manner the great poet Maha Kavi Barathiyar also said that if there is no food available for even a single person then we should destroy the whole world. The desire and anger of both the poets are only towards the society. The society or the government should fulfill the basic needs of the common people. The prime goal of the society is to provide basic amenities to the people.

In the period of Barathiyar and Thiruvalluvar, food scarcity was one of the main problems. Hence, they were concerned about this issue. However drinking water and breathing air might have been available in purest form on those days. But now the availability of pure air and safe water are also becoming a big problem. Water, air, soil and everything are getting spoiled due to human anthropogenic development activities.

If we deeply analyse the statement of both poets, it concludes that the available natural resources in the world are common and equal for each and every person in this world. Everyone should get access to all the natural resources, basic amenities and basic needs without any interruption. If they do not get the same then it is not their fault and it is the responsibility of the society or the ruler. The society should correct its mistake and provide the basic amenities equally to each and everyone. If Thiruvalluvar and Barathiyar are alive today they would curse the entire world for not providing clean air and clean water for everyone.

Let us see how the common man gets affected due to present development and starved of basic needs.

All over the world every individual is not consuming the same amount of natural resources. The per capita natural resource consumption will vary depending upon the countries and their development. With respect to natural resource (oil, gas, coal etc) consumption, the emissions of CO_2 quantity per capita also varies. Country wise data of CO_2 emission per capita is given below.

Country	CO_2 Emissions Per Capita (Metric Ton/Year)
United State	16.5
Australia	15.4
Canada	15.1
Netherlands	9.9
Japan	9.5
China	7.5
India	1.7

When we take a look at the above data, in United states, a single person is emitting 16.5 metric tons of CO_2 per year. At the same time per capita CO_2 emissions per year in India is only 1.7 metric tons. So per capita CO_2 emission by a common man in US is almost 10 times more than that of an Indian. In developed countries like USA , people are consuming more power and lead a luxurious life, on the other hand people in developing countries like India are consuming less power and emitting less CO_2. However the adverse effects of CO_2 in the atmosphere like global warming and acid rain commonly affect all the people irrespective of developed or under developed countries. Even within the same country some will be leading a very luxurious life by exploiting more natural resources and the poor in the same country will be leading a simple life even without a single two wheeler. But the effects of pollution are common to everyone irrespective of poor or rich. Due to the developmental activities, the innocent poor and common man should not be affected for getting his basic needs such as air, water and food. This is the DHARMA of development. The rich man who is grabbing more natural resources for his luxurious life should have more responsibility for clearing the polluted atmosphere. In that way, the developed countries must take more responsibility for bringing down the concentration of greenhouse gases in the atmosphere.

It is appropriate to remember the warnings of both Thiruvalluvar and Barathiyar here, they told a common man should not suffer due to any development. They both vented out their anger towards the rulers of the society.

Air Pollution:

Clean, dry air consists primarily of nitrogen and oxygen 78 percent and 21 percent respectively, by volume. The remaining 1 percent is mixture of other gases, mostly argon (0.9 percent), water vapour along with traces (very small) amounts of CO_2, methane, hydrogen and helium. If this air composition is disturbed - increased or decreased by any means, it can be harmful to the living organisms. So many hazardous gases are accumulating in the atmosphere from various industrial sectors and also from the agricultural sector. But important observation here is the oxygen concentration in the atmosphere should not decrease below 19.5%. If it does, then life is not possible in the earth.

There is a relation between oxygen and CO_2 concentration in the atmosphere. Whenever CO_2 concentration is increased in the atmosphere then the same amount of oxygen will decrease.

$$C + O_2 \longrightarrow CO_2$$

If one mole of CO_2 increases in the atmosphere then the same one mole of oxygen gets depleted from the atmosphere. Increasing CO_2 concentration also gives more dangerous effects to the environment. Our aim should be focused towards bringing down the CO_2 concentration level and building up the oxygen concentration level in the atmosphere.

Let us calculate the quantity of oxygen depletion in the atmosphere by burning of fossil fuel.

Total weight of oxygen in the atmosphere $= 1080000 \times 10^9$ ton

Fossil fuel burning per year is 5.809×10^9 ton

For burning 12 tons of carbon, 32 tons of oxygen is required and 44 ton of CO_2 will be emitted to the atmosphere

$$\text{CO}_2 \text{ emission per year} = \frac{44.0}{12.0} \times 5.809 \times 10^9 \text{ ton}$$

$$= 21.3 \times 10^9 \text{ ton}$$

(If 44 ton of CO_2 emitted 32 ton of O_2 will be depleted from the atmosphere)

Oxygen depletion per year:

$$= \frac{32.0}{44.0} \times 21.3 \times 10^9$$

Oxygen depletion per year = 15.49×10^9 ton

Concentration of O_2 depletion in the atmosphere per year

(Total weight of oxygen in the atmosphere 1080000×10^9 ton)

$$= \frac{15.49 \times 10^9}{1080000 \times 10^9} \times 10^6$$

$$= 14.3 \text{ ppm}$$

If the current rate of fossil burning is not controlled, then the atmospheric oxygen concentration will come down to 14.3 ppm. per year

Current level of oxygen in the atmosphere is 20.946%v/v

Minimum oxygen concentration required for leading comfortable life as per OSHA standard is 19.5%

So the difference of concentration between the current value and the OSHA standard is 20.946 − 19.5 = 1.446%

The time period required for depletion of oxygen to 1.446% in the atmosphere

$$= \frac{1}{0.001434} \times 1.446$$

$$= 1008 \text{ years}$$

As per this calculation, if we are not taking proper steps to control current usage of fossil fuels, then O_2 concentration will be depleted below the minimum level within 1008 years. Which means as told earlier in this chapter, life will not be possible after 1008 years.

In pre industrial period the CO_2 concentration was 275 PPM. At present the carbon dioxide concentration in the atmosphere is more than 400 PPM. Depleting oxygen concentration in the atmosphere will give more adverse effect than the increasing CO_2 level. Increasing CO_2 concentration leads to global warming and decreasing oxygen concentration will lead to extinction of bio organisms in the earth.

If this trend of increasing CO_2 concentration in the atmosphere continues further, then the next generation will be definitely leading their life only with the support of oxygen cylinders. In future, there will be a huge market for oxygen sales. Every individual will carry oxygen cylinder on their backs for breathing. Unfortunately this might be the deplorable fate of our future generation. Now we have to take stringent action against CO_2 emissions.

Carbon Intensity:

Carbon intensity is an important parameter which indicate the quantity of CO_2 emitted to the atmosphere for producing one unit of power (Kwhr) by various processes. This quantity will vary depending upon the nature of fuel that used for power production.

Some of the values of carbon intensity are given below for each process.

Nature of Power	Carbon Intensity
Hydroelectric power	4.0
Wind energy	12.0
Nuclear energy	16.0
Solar energy	22.0
Geothermal energy	45.0
Natural gas	469.0
Coal	1001.0

From the above data we can easily conclude which type of the power production process is eco friendly nature. Renewable energy like solar energy, wind energy and hydroelectric power have very low value of carbon intensity, whereas natural gas and coal are having more carbon intensity value. Carbon intensity value for coal based thermal power plant is approximately hundred times more than that of wind energy. Plenty of energy is available in renewable resources like hydro power, wind energy, solar energy and geothermal energy. Even nuclear energy has a low level of carbon intensity value. We have to find new technologies to harvest more energy from the renewable energy sources, thereby completely eradicating thermal Power.

செயத்தக்க அல்ல செயக்கெடும் செயத்தக்க
செய்யாமை யானும் கெடும்.

To do what ought not to be done is dreadful:
Not to do what ought to be done is equally so.

இந்தக் குறளுக்கு ஏற்ப செய்யத் தகாத செயலைச் செய்வதாலும் கெடும், அதேபோன்று செய்ய வேண்டிய செயலைச் செய்யாமல் விட்டாலும் கெடும். தற்பொழுது செய்யத்தகாத செயல் கரிம எரிபொருள்களைப் பயன்படுத்தி மின்சாரம் தயாரித்தல். இதனால் வாயு மண்டலம், நீர் மண்டலம் மற்றும் நிலம் ஆகிய மூன்றும் சீர்கேடு அடைகின்றன. செய்ய வேண்டிய செயல் புதுப்பிக்கத்தக்க ஆற்றல் மூலங்கள் வழியாக மின்சாரம் தயாரிக்கும் முறைகளைக் கையாண்டு மின்சார உற்பத்தியைப் பெருக்கி சுற்றுப்புற சுகாதாரத்தைப் பேணிக் காக்க வேண்டும்.

As stated in this kural if we do the things which should not be done, we will get to suffer and if we don't do the things which should be done also, we will get to suffer. Therefore, we should not produce power by using fossil fuels. But we are doing so and spoiling our environment. We should only use the renewable energy for producing power and save our environment.

"God is like oxygen, you can't see him
but you can't live without him"

8 NATURAL BOUNDARIES OF THE COUNTRY

இருபுனலும் வாய்ந்த மலையும் வருபுனலும்
வல்லரனும் நாட்டிற்கு உறுப்பு.

Lakes above and wells below, mountains well located,
Rivers and streams flowing from them and strong fortresses
establish a country.

ஊற்று நீரும், ஆற்று நீரும், இயற்கையாக அமைந்த மலையும், மலையில் உண்டான காடுகளும் ஒரு நாட்டிற்கு இருக்கவேண்டிய முக்கியமான அம்சங்களாகும். நீரைப் பற்றிக் கூறும் பொழுது ஊற்று நீர் மற்றும் ஆற்றுநீரை முக்கியமாக குறிப்பிடுகின்றார் திருவள்ளுவர். மழை பெய்தால் ஆறுகளில் தடையின்றி தண்ணீர் வரும். அந்தத் தண்ணீரை ஏரி, குளம், குட்டை, கண்மாய், ஊறணி, ஊருணி, கேணி, தடாகம், பொய்கை மற்றும் கயம் போன்றவற்றில் சேமித்து ஊற்று நீரை அதாவது நிலத்தடி நீரை மேம்படுத்த வேண்டும். நிலத்தடி நீர் மேம்படுத்தப்பட்டால் அந்த நாட்டில் விவசாயம் செழித்தோங்கும். மேலும் மழை பெய்யும் காலங்களில் மழை நீர் எந்த தடையும் இல்லாமல் பூமியில் இறங்கி நிலத்தடி நீரின் அளவை மேம்படுத்தச் செய்ய வேண்டும். இயற்கையாகப் பெய்யும் மழை நீரின் அளவு ஒவ்வொரு வருடமும் ஏறக்குறைய போதுமான அளவாகவே இருக்கிறது. நீர்நிலைகளைச் சரி செய்தாலே நமக்கு வருடத்திற்குப் போதுமான நீரைச் சேமித்து வைத்துக் கொள்ளலாம்.

திருவள்ளுவர் மேற்கண்ட குறளில் மலை மற்றும் காடுகளின் சிறப்புகளையும் விளக்கியுள்ளார், எந்த ஒரு நாடு காடுகளின் வளத்தையும் மற்றும் பரப்பளவையும் சரியாக பராமரிக்கிறதோ அந்த நாட்டில் எல்லா வளங்களும் சிறப்பாக அமையும். இப்பொழுது நம் மனித வளர்ச்சிக்காக நகரியம் மற்றும் தொழிற்சாலைகளை அமைப்பதற்காகக் காடுகளை அழித்துக் கொண்டிருக்கிறோம். இதைத் தவிர்க்க வேண்டும் என்பதற்காகவே திருவள்ளுவர் இக்குறளின் மூலம் காடுகளின் முக்கியத்துவத்தை வலியுறுத்திக் கூறியுள்ளார்.

Thiruvalluvar explained in this kural about the importance of ground water, river water, mountains and forests. These four are the main resources for the countries development and its richness.

Rivers:

Rivers are the backbone of human civilization. They provide us fresh water which is helpful for various purposes such as agriculture, industries, drinking and other domestic usages. Without rivers, life will come to a halt. We humans exist because of rivers. Importance of Rivers cannot be stated in just few words.

Prominent Uses of Rivers

1. Rivers are major source of fresh water. Around 96% of the Earth's water body consists of saline water that exist in the oceans which cannot be consumed by humans. Hence we completely depend on rivers for drinking and other essential uses. If we don't get good amount of water flow through the rivers, then life on earth will come to an end.

2. There are various civilization formed on the banks of the rivers. Some of the earliest ones include the Nile Valley civilization, Indus Valley civilization, Yellow River Valley civilization etc. These civilizations started near rivers because river plains had fertile soil which helps in cultivation. Moreover they also helped in transportation. Even today, many villages and cities are based near rivers.

3. Rivers are not only important for human beings but also serves a great purpose to animals and trees as well. There are various aquatic animals which breed in rivers. Moreover, various plants grow in the river beds. They form an essential part of the ecosystem which is the most important to maintain the balance in the food chain.

4. Rivers are also a source of energy. It helps in generating electricity. In the hilly areas, rivers have a lot of current in it. This energy can be converted into electricity. Hydroelectric plants are built across the rivers for generating electricity. Various dams are built for producing electricity. The average elevation of land from

the sea level is approximately 850 meters. Due to this huge elevation difference between land level and sea level, we can harvest sufficient energy from the hydroelectric plants.

The Ganga or popularly known as The Ganges, is undoubtedly India's most sacred place but sadly it isn't sacred anymore. What used to be a place to wash your sins away is now become a place where sins happen. More than half of India's population is dependent on the waters of river Ganga. But the immersion of ashes and dead bodies in the river is one of the primary causes of pollution of the Ganga.

It is very important to keep the river path clean in order to ensure good flow of water and without any disturbances like encroachment and mining of sand. It can also show its fierce face as floods when the flow exceeds the limits due to abundant rain or encroachment or by destroying the water bodies like ponds, lakes etc. It takes away a large number of lives and destroy huge amount of properties when it floods. So we should not create any disturbance in the river's path. We should keep them clean and let them flow steadily. We will never know the worth of rivers until they dry up. A drop of water is worth more than a sack of gold to a thirsty man. We should understand the importance of the rivers and save them from pollution.

IMPORTANCE OF GROUND WATER

Groundwater represents about 30% of world's fresh water. Nearly 69% of fresh water is locked up in the ice caps and mountain snow/glaciers. Merely 1% of the fresh water is found in river and lakes. Groundwater is a very important resource for human usage. Almost one third of the fresh water is being consumed only from the ground water source and at some parts of the world it is 100%.

Groundwater has a significant role in the economy too. It is the main source of water for irrigation and for the food industry. In general, groundwter is a reliable source of water for agriculture and can be used in a flexible manner. when the area gets affected by drought, more groundwater will be extracted and when the rain falls the groundwater will be recharged again. It is estimated that nearly 43% of ground water is used for irrigational purposes.

Groundwater is found everywhere and its quality is usually very good. Groundwater is stored in the layers beneath the surface and sometime at very high depths, helps protecting it from contamination and preserve its quality. Additionally, groundwater is a natural resource which can often be found close to the final consumers and therefore does not require large investments in terms of infrastructure and treatment. The most important fact about using groundwater is to find the right balance between withdrawing and letting the aquifer's level to recover to avoid over exploitation and to avoid pollution of this crucial resource.

Now a days concrete roads, buildings, urbanisation and plastics wastes are blocking the percolation of water into the ground and hence ground water level decreases rapidly.

Importance of Forests

Production of Oxygen:

Even though all civilizations were developed in the bank of the rivers, forests are the origin for life. They not only provides basic needs such as food, shelter and dress, they are main source for oxygen and water. All animals of the entire planet depends only on forests for their survival.

The most important function of forests is producing huge amounts of oxygen for entire biosphere in the earth as a by-product of photosynthesis. Oxygen is the main respiratory gas for all animals, it ensures our survival. During photosynthesis, trees are not only producing oxygen, they also absorb CO_2 from the air. CO_2 is one of the main pollutants of air now. Forests prevent soil erosion and also play an important role in the water cycle and control moisture levels in the atmosphere. And finally, forests are the natural home and habitat for millions of species of animals, birds, and insects.

The early atmosphere was made up mainly of CO_2 and water vapour. Water vapour condensed to form the oceans. Photosynthesis causes decrease of carbon dioxide and increase oxygen About 2.2 to 2.7 billion years ago, the early bacteria known as cyanobacteria is the only main source of oxygen in the atmosphere.

A single mature leafy tree will be producing enough oxygen for 2–10 people. Plants in the land maintain oxygen level in the atmosphere. Likewise plants in ocean maintain the O_2 level in the hydrosphere.

Total production of oxygen from land photosynthesis is

$$16500 \times 10^{10} \text{ kg per year}$$

Total production of oxygen from ocean photosynthesis is

$$13500 \times 10^{10} \text{ kg per year}$$

Water Cycle:

Forests help in maintaining water cycle on earth. Plants absorb water from the soil through their roots. The process of releasing excess water by plants into the atmosphere in the form of water vapor is known as transpiration. Forests play a significant role in continuing water cycle. Plants keep only 5% of the total water that absorbed through the root, remaining 95% of the water is let into atmosphere by the process of transpiration.

In such a way plants suck huge amount of the ground water free of cost and add them into the water cycle. Otherwise the ground water will be locked up as dead investment in underground.

Shelter, Food Employment and Medicine:

Forest is the home to many living organisms. It is one of the precious resources provided by nature. The organisms living in forests are interdependent on each other. 300 million people live in forests, including 60 million indigenous people.

The average forest land should be above 33% of total land area of the country. The area of land covered by forest is a key piece of information for a healthy nation. Total forest area in 2005 was estimated around 30% of the planet's land area, just under 40 million km^2. This corresponds to an average of 0.62 ha (6200 m^2) per capita, though it is unevenly distributed.

South America is the only region with the highest percentage of forest cover (almost half of the land area) and Asia is the region with the lowest percentage of forest cover (less than 20% of land area). In India the total forest area is only 22%. National forest policy had a target of keeping 33% of land area as forest since it is the minimum requirement needed for maintaining ecological balance in a country. Forest conservation denotes the preservation, afforestation and the protection of forests. It also involves the reversal of deforestation and environmental pollution. The preservation of forest resources is absolutely essential for maintaining good balance in ecosystem.

MOUNTAINS

Mountains play a significant role in providing water and food supply to millions of people in the world. Mountains cover around 22% of the surface of the earth and 13% of the world's population live in the mountains.

Mountains directly support 22% of the world's people who live within mountain regions. Lowland people also depend on mountains for a wide range of goods and services, including water, timber, biodiversity maintenance, recreation and spiritual purposes. Mountains provide freshwater for more than half of humanity. Mountains are considered as the water towers of the world. Mountains offer 60–80% of the world's freshwater. According to UNO, almost one billion people live in mountain areas, and over half of the human population depends on mountains for water and nutrition. Mountains not only provides essential things, it also provide energy like Hydro electric power. The height of the mountain provides potential energy to the water, which is harvested in the hydroelectric power plants.

Thiruvalluvar not only described the rivers, mountains and forest as natural resources, he also mentioned them as a natural barrier to the country. It will protect the country from the invasion of foreign powers. So we should not wantonly destroy our natural resources for anything.

மணிநீரும் மண்ணும் மலையும் அணிநிழற்
காடும் உடைய தரண்

Pearl like perennial water, sand, mountains and shaded trees
A strong fortress contain all these.

ஒரு நாட்டிற்குத் தெளிந்த தூய்மையான நீர்நிலையும், அகன்ற விவசாயம் செய்வதற்கு ஏற்ற நிலப்பரப்பும், உயர்ந்த மலையும், அடர்ந்த காடுகளும் அரண்களாகும். இவையே நீரரண், நிலவரண், மலையரண் மற்றும் காட்டரண் என்னும் இயற்கை அரண்கள் ஆகும். இந்த நான்கு அரண்களையும் நாம் பாதுகாத்தால் எல்லா இயற்கை வளங்களும் தடையின்றி நமக்குக் கிடைக்கும். இதைத்தான் நமது பொய்யாமொழிப் புலவர் வள்ளுவர் திருமகனார் மேற்கண்ட குறளின் மூலம் இயற்கை வளங்களின் அவசியத்தையும், அவற்றைப் பாதுகாப்பதன் முக்கியத்துவத்தையும் வலியுறுத்திக் கூறியுள்ளார்.

Rivers are water barriers, whereas forests and mountains are land barriers. In this way ancient kingdoms protected themselves from the enemy.

> "The death of the forest is the end of our life".
>
> – Dorothy Stang

9 IMPORTANCE OF ATMOSPHERE

வான்நின்று உலகம் வழங்கி வருதலால்
தான்அமிழ்தம் என்றுஉணரற் பாற்று.

The world lives because of the rains and hence
Rain is known to be the nectar of immortality.

திருவள்ளுவர், வானத்தில் இருந்துதான் மனித உயிர்களுக்கும், மற்ற எல்லா உயிர்களுக்கும் உயிர் வாழ்வதற்குத் தேவையான எல்லா அடிப்படையான காரணிகளும் கிடைக்கின்றன என்று இந்தக் குறளின் மூலம் கூறுகின்றார். உதாரணமாக ஒளி, நீர், வெப்பம், காற்று இவை அனைத்துமே வானத்தில் இருந்து தான் கிடைக்கின்றன. எனவே நாம் வானத்தை எந்தவிதமான மாசுபடாமல் தூய்மையாக வைத்திருக்க வேண்டுமென்று 2000 ஆண்டுகளுக்கு முன்பே மனித குலத்தின் நன்மைக்காக மிகவும் வலியுறுத்திக் கூறிச் சென்றுள்ளார். ஐம்பூதங்களில் நிலம் மட்டுமே பூமியில் இருக்கின்றது. மற்ற நான்கும் வானத்தில் தான் இருக்கின்றன. நாம்ஆகாயத்தை மாசுபடுத்தினால் பூமியும் மாசுபட்டு அதன் தன்மையை இழந்துவிடும்.

தற்பொழுது நாம் அறிவியலின் வளர்ச்சியால் பல தொழில்நுட்பத்தைக் கையாண்டு மனிதகுல வளர்ச்சிக்காகவும், மேம்பட்ட நாகரீக முன்னேற்றத்திற்காகவும் ஆகாயத்தைப் பலவழிகளில் சீர்கேடு செய்து கொண்டிருக்கிறோம். அனல்மின் நிலையங்களின் மூலம் மிகுதியான கார்பன் டை ஆக்சைடு, சல்பர் டை ஆக்சைடு, மற்றும் நைட்ரஜன் ஆக்சைடுகள் போன்றவற்றையும் விண்ணில் செலுத்திக் கொண்டிருக்கிறோம். மேலும் குளிர்விப்பானில் பயன்படுத்தும் குளோரோ புளோரோ கார்பன் போன்றவற்றை மிகுதியான அளவு வான்வெளியில் செலுத்தி ஆகாயத்தை தூய்மையற்றதாக மாற்றிக் கொண்டிருக்கிறோம். இவற்றை நாம் உடனடியாக நிறுத்தாவிட்டால் இந்த பூமியே மனிதகுல வாழ்விற்கும் மற்ற ஏனைய உயிர்களின் வாழ்விற்கும், இன்னும் அறுதியிட்டுக் கூற வேண்டுமானால் தாவர வளர்ச்சிக்கும் கூட ஏற்றதற்றாகிவிடும்.

Thiruvalluvar described in this kural that all essential resources for our life such as light, heat, water and air are received from the sky. Four among the five pancha boothas except soil (land) are coming only from the sky. If we spoil the sky, then all the things in the earth will become polluted and the entire environment will not be suitable for living things in future.

All these pancha boodhas are getting affected in the sky by human anthropogenic activities.

Water (நீர்)

For entire bio organisms that lives in earth, sufficient quantity of water is being received as rain. Rain water is the purest form of water. Nature is doing mega distillation with the help of solar heat.

However this purest rain water is being polluted in the sky by gases like sulfur dioxide and nitrogen oxides, emitted by human anthropogenic activities. Hence the water become acidic even before reaching the surface of the earth. The rain become acid rain in the sky itself. Nature provides sufficient quantity of pure water to the wellness of each and every bio organism. If we properly receive, store and utilize even only 10% of rain water, that is more than sufficient for the entire world. But we are spoiling the quality as well as the quantity of rain water and release most of the rain water to ocean without proper storage.

Air (காற்று)

Air is directly getting polluted due to the venting of more hazardous gases into the atmosphere from industries and vehicles. Fossil fuels are not only emitting gases, they also emits heavy metal like lead (pb), mercury (Hg), cadmium (Cd), and copper (cu). These heavy metals reach the earth surface through rain and contaminate the land surface.

Heavy metals that are present in the fossil fuels are

Lead   7.0 grams per ton

Mercury   0.05 grams per ton

Cadmium	0.5 grams per ton
Copper	10.0 grams per ton

These heavy metals are highly poisonous to the living things. They reach the ground through rain water. In this way, land is getting degraded due to pollution in the atmosphere (sky).

Fire (நெருப்பு)

Fire means heat. The earth surface temperature is getting affected by the greenhouse gases. Greenhouse gases like carbon dioxide (CO_2), methane (CH_4), water vapour (H_2O), etc. are blocking the outgoing radiation from the earth and increases the earth surface temperature. So earth surface temperature is being affected by the pollution accumulated in the sky.

Light (ஒளி)

Light is very important component in the environment. Without light plants cannot produce food by the process of photosynthesis. From the solar radiation visible rays, IR rays and UV rays are coming towards the earth. Among all, the UV rays are highly harmful to living things including plants.

Naturally the ozone layer available in the stratosphere region filters the harmful ultraviolet rays. Hence ozone layer act as a protector of earth surface. But now this ozone layer is being destroyed by the CFC (chlorofluorocarbons) which are emitted by the refrigerators, air conditioners, etc.

Hence we can say that the light (ஆகாயம்/Sky) is also getting polluted by more quantity of UV rays in the sky.

Land (நிலம்)

If we protect the sky (atmosphere) from pollution then automatically land will be rejuvenated

As per Thiruvalluvar's vision, if we keep the sky clean, then the entire surface in the earth will have clean environment. All the living things in the earth will have sufficient quantity of pure rain water, fresh air and clean environment.

மேலே கூறிய குறளை மறுபடியும் நினைவுகூர்ந்தால் திருவள்ளுவர் கூறிய கூற்றின் உண்மை நமக்கு நன்கு புரியும்.

If we recall the kural again, we ll understand the real meaning of the great poet's statement.

வான்நின்று உலகம் வழங்கி வருதால்
தான்அமிழ்தம் என்றுஉணரற் பாற்று

நாம் பூமியில் வாழ்ந்தாலும் நமக்கு எல்லா விதமான அத்தியாவசியமான பொருள்கள் அனைத்தும் வானத்தில் இருந்துதான் கிடைக்கின்றன. எனவே நாம் வானத்தைத் தூய்மையாக வைத்திருந்தாலே போதும், நமக்குக் கிடைக்க வேண்டிய அனைத்தும் தடையில்லாமல் கிடைத்துக் கொண்டிருக்கும். நாம் வானத்தை எந்த அளவிற்கு மாசுபடுத்துகிறோமோ அந்த அளவிற்கு பூமியும் மாசுபடும் என்பதை நாம் எல்லோரும் உணர வேண்டும் என்று இரண்டாயிரம் ஆண்டுகளுக்கு முன்பே திருவள்ளுவர் கூறிவிட்டு சென்றிருக்கிறார். இதை நாம் மனதில் கொண்டு கூடுமானவரை வானத்தைத் தூய்மையாக வைத்திருக்கவேண்டும்.

Though we live in the land, we receive all the essential resources from the sky. So if we keep our sky clean, we will get everything we need. If we spoil the sky by polluting it, then our earth also will get polluted in the same way. The great poet had a vision and stated this some 2000 years ago. So we should keep this in our mind and try to keep the sky clean.

Conclusion:

அழிவதூஉம் ஆவதூஉம் ஆகி வழிபயக்கும்
ஊதியமும் சூழ்ந்து செயல்.

What gets wasted, what gets created and the returns
The creation yields, consider all these before starting a task.

ஒரு செயலைத் தொடங்குமுன் அதனால் ஏற்படும் அழிவையும், அழிந்த பின் ஆவதையும், பின்பு உண்டாகும் நன்மையையும் ஆராய்ந்து செய்ய வேண்டும்.

While starting any development activities for the uplift of the mankind, we should analyse both the beneficial effects and adverse effects of that activity. Finally after considering the benefits, all the pros and cons and making sure that there aren't much adverse effects, we should proceed to action.

> "The earth, the air, the land and the water are not an inheritance from our forefathers but on loan from our children. So we have to return them at least as it was handed over to us."
>
> — Mahatma Gandhi

10 GOOD AND BAD EFFECTS OF UV–RAYS, METHANE AND LIGHTNING

குணம்நாடி குற்றமும் நாடி அவற்றுள்
மிகைநாடி மிக்க கொளல்.

Weigh upon the virtues and flaws, discern which is more,
And thus decide.

எந்த ஒரு பொருளும் நூறு விழுக்காடு நன்மையைத் தராது. அதேபோன்று நூறு விழுக்காடு தீங்கையும் தராது. ஒரு பொருள் காலத்துக்குத் தகுந்தவாறும் இடத்திற்குத் தகுந்தவாறும் நன்மை தீமைகளைக் கொடுக்கும். நாம்தான் பொருளின் தன்மையை ஆராய்ந்து இடம், பொருள் அறிந்து அதனைப் பயன்படுத்திக் கொள்ள வேண்டும். ஓடாத கைக்கடிகாரம் கூட ஒரு நாளைக்கு இரண்டு முறை சரியான நேரத்தைக் காட்டுகிறது. எனவே எந்தப் பொருளையும் முழுவதும் தீமையானது என்று ஒதுக்கிவிட முடியாது. இதைத்தான் திருவள்ளுவர் ஒரு பொருளின் குணம், குற்றம் இவை இரண்டையும் ஆராய்ந்து அதில் எது மிகுதியானதோ அதை எடுத்துக்கொள்ள வேண்டுமென்று இக்குறளின் மூலம் தெளிவுபடுத்துகிறார்.

Nothing can be disposed as waste in the universe. Even a broken clock is right twice a day. Everything has its own value. It differs depending on the place, time and condition. We can get both beneficial as well as ill effects from the same object. We have to analyze things completely and then only we can come to a conclusion about whether it can be used or discarded for a particular purpose. So we should not make final decision before evaluation.

Let us correlate this with some examples in environmental science.

1. GOOD AND BAD EFFECTS OF UV RAYS

Most people know that the ultraviolet rays of sun gives adverse effects to plants and animals and so it is very harmful in nature but many of us does not aware that they also gives good effects that plant and animals need.

BAD EFFECTS OF UV – RAYS

Skin Cancer – UV is an environmental human carcinogen. It is the most prominent and universal cancer-causing agent in our environment. There is very strong evidence that each of the three main types of skin cancers are caused due to exposure of UV rays. Research shows that as many as 90% of skin cancers are due to UV radiation.

Sunburn – Sunburn is a burn that occurs when skin cells are damaged due to the absorption of energy from UV rays. Skin turns red when sunburn appears.

Damage on Immune System – Over-exposure to UV radiation has a harmful suppressing effect on the immune system. Exposure to UV rays can change the distribution and function of disease-fighting white blood cells in humans within 24 hours. Repeated over-exposure to UV radiation can cause even more damage to the body's immune system. The immune system defends the body against bacteria, microbes, viruses, toxins and parasites .

Effects on Eyes – Prolonged exposure to UV damages the tissues of eyes and can cause a 'burning' of the eye surface, called snow blindness. Normally this effect will disappear in couple of days, but may lead to further complications later in life.

Aged Skin – UV speeds up the aging of skin since it destroys tissues beneath the top layer of the skin. This causes wrinkles, brown 'liver' spots and loss of skin elasticity.

Bad Effects of UV Radiation on Plants:

Plants are highly sensitive to the UV radiation. In plants, UV radiation damages the cell membrane and internal parts. UV rays are directly or indirectly affecting the metabolic process such as photosynthesis, respiration, growth and reproduction. Finally UV rays damage plant crop and reduce the yield and quality.

Now let us discuss the good effects of UV rays.

GOOD EFFECTS OF UV RAYS

Vitamin D Synthesis:

UV rays are very much essential for synthesis of Vitamin D. Sufficient quantity of Vitamin D in human body is made through a series of biochemical processes that start when the skin is exposed to the sun's UV rays. Though Vitamin D is available naturally in fish and eggs, its contribution level is only 5 to 10% to the body's overall vitamin D requirement.

Importance of Vitamin D:

Vitamin D controls calcium levels in the blood. It is needed to develop and maintain healthy bones, muscles and teeth and is also important for general health. Vitamin D is crucial for bone and muscle development. The recommended vitamin D levels are 60–70 nano mole/L. Deficiency of vitamin-D will increase risk of bowel cancer, heart disease, infections, auto-immune diseases and rickets in children. In addition, it has been identified that vitamin D has a positive effect on controlling high blood pressure. Blood pressure will be maintained normal when vitamin D levels are normal. Furthermore, an inadequate level of vitamin D is a direct contributing factor for type 1 diabetes, rheumatoid arthritis, autoimmune disease of the thyroid and inflammatory bowel disease. This much vital nutrient that required for good health can only be received through UV radiation.

2. OTHER BENEFICIAL EFFECTS OF UV-RAYS

Useful for disinfection and sterilisation – UV has positive applications in the fields of disinfection and sterilisation. UV can effectively 'kill' (deactivate or destroy) microorganisms such as viruses and bacteria. To destroy the microorganisms, UV rays penetrate the cell's membrane, destroying the DNA, and stops its ability to reproduce and multiply. This destructive effect explains why we should use UV antibacterial lamps for disinfection and sterilisation.

Thus the UV radiation has both beneficial and non beneficial effects, we have to analyse and come to the final conclusion. More exposure of body in UV rays will create the risk of skin cancer and skin disease. Less exposure or no exposure in UV rays reduce the synthesis of Vitamin D which is more important nutrition for human health.

The ozone umbrella filters the great proportion of ultraviolet rays of the sunlight. If all the ultraviolet rays reach the surface of the earth, we would be sunburned to death. However, only a small amount of ultraviolet rays reaches the Earth which helps to synthesis vitamin D. If the ozone layer is just twice concentrated as it is now, even a minimum amount of ultraviolet rays will not reach the earth and there will be great lack of Vitamin D, making us to suffer from the disease like rickets and other health problems.

(Too much exposure to UV radiation increase the risk of skin cancer and no exposure or less exposure reduce the synthesis of vitamin-D. How much UV exposure a person needs, depends on the season UV levels, skin type and current vitamin D levels.)

NOW LET US DISCUSS THE GOOD AND BAD EFFECTS OF METHANE

Bad Effects of Methane:

Methane is a highly potent green house gas. Its global warming potential is 56, it means the greenhouse effect of methane is 56 times greater than CO_2. For example, one kg of methane emission into the atmosphere is equal to 56 kg of CO_2 emission. Accumulation of methane in the atmosphere is highly dangerous as it leads to global warming.

Good Effects of Methane:

Methane plays a significant role in saving the ozone layer. It prevents the depletion of ozone layer in the stratosphere.

Chlorofluorocarbons are major source of ozone destroying agents. We are using CFC in the refrigerators. Chlorofluorocarbons are stable in nature, they travel slowly to the stratosphere region then it will be destroyed there by the UV rays and frees the chlorine atom from the CFC molecule as follows,

$$CF_3Cl_3 \longrightarrow CF_3Cl_2 + Cl$$
$$\text{(Uv rays)}$$

This freed Cl atom will destroy ozone molecule in the stratosphere region as follows

$$O_3 + Cl \longrightarrow ClO + O_2$$

The ClO will react with another ozone and destroy it again and make free one chlorine atom.

$$ClO + O_3 \longrightarrow 2O_2 + Cl$$

This freed chlorine atom again will react with another ozone molecule, like that one chlorine atom will stay more than 200 years in the atmosphere and spoils more than one lake ozone molecules approximately. The freed chlorine atom can not be easily destroyed.

But here the methane molecule available in the atmosphere is destroying the chlorine free radicals. Methane react with active chlorine atom and form HCl. This HCl is dissolved in water during rainfall and drained out to the earth surface.

$$CH_4 + Cl \longrightarrow HCl + CH_3.$$

Like methane NO_2 also react with Clo and form $ClONO_2$, a stable molecule. In that way the active chlorine atom in the stratosphere region is removed by CH_4 and NO_2.

In one way methane present in the atmosphere leads to global warming and on the other hand the same methane is used to prevent the ozone layer.

GOOD AND BAD EFFECTS OF LIGHTNING

Lightning also has good and bad effects.

DANGEROUS EFFECTS OF LIGHTNING

A lightning strike generates a huge amount of electric charge and also has a temperature of about 30000 degree Celsius.

It can cause the following adverse effects to humans, property and environment.

1. Lightning can cause fire and damage in buildings, resulting in more destruction and damage to the property especially to electrical and electronic equipment.

2. Lightning can cause forest fires.

3. Lightning cause serious injuries or sometimes even kills the animals and human beings instantly.

The Impact of a Lightning Strike

About 2,000 people have been killed worldwide by lightning each year. Lightning strikes can cause cardiac arrest and severe burns. Hundreds of people who survive after lightning strikes later suffer from a variety of lasting symptoms, including memory loss, dizziness, weakness, numbness and other life-altering ailments.

In this way lightning produces very dangerous adverse effects to the environment, property and human beings.

POSITIVE IMPACT OF LIGHTNING

Rain without lightning and thunder is only gain of water from the atmosphere. But rain with thunder and lightning is adding very crucial nutrient of nitrate to the soil. People in olden days know the secret of this. They described the rain without thunder in Tamil as "மலட்டு மழை" (impotent rain). Sometimes we also observe the blooming of mushroom in the morning in village side after heavy thunderstrom rain in the night.

In spite of all adverse effects, lightning also gives beneficial effects in the field of agriculture by improving soil nitrogen content. Nitrogen content in the soil is very much important for the crops to achieve optimum yields. Nitrogen is so vital because it is a major component of chlorophyll. It is also a major component of amino acids, the building blocks of proteins. Without nitrogen, there would be no life as we know it.

EFFECTS OF LIGHTNING IN THE SOIL

The atmosphere is having 78 percent nitrogen. The two atoms in the nitrogen molecule are held together very tightly and cannot be separated easily. Nitrogen in the atmosphere is highly stable and nonreactive due to the triple bond between atoms in the nitrogen molecule. Even though plenty of nitrogen is available in the atmosphere, plants cannot make use of it.

Nitrogen is added to the soil by two natural processes. One process is bacterial activity – Organic Nitrogen that's present in soil organic matter, crop residues and manure is converted to inorganic Nitrogen through the mineralization process. In this process, bacteria digest organic material and release NH_4^+-N.

Formation of NH_4^+-N increases as microbial activity increases.

Lightning is another natural way. Nitrogen in the atmosphere can be transformed into the soil by the lightning. Each strike of lightning carries electrical energy that is powerful enough to break the strong bonds of the nitrogen molecule in the atmosphere. Once it splits, the nitrogen atoms quickly combine with oxygen in the atmosphere forming nitrogen dioxide. This nitrogen dioxide dissolves in water, creating nitric acid, which forms nitrates. The nitrates reach to the ground through raindrops and seep into the soil making it very fertile.

It is estimated that a flash of lightning produces 4×1026 molecules of NOx,. The quantity of nitrogen added to the soil by this thunderstorm activity is estimated to be 20 lb nitrogen per acre per year.

நன்மையையும் தீமையும் நாடி நலம்புரிந்த
தன்மையான் ஆளப் படும்.

A person, because of his ability to perform well by
assessing pros and cons, Will assume charge of
key assignments.

ஒரு பொருளின் நன்மையையும், தீமையையும் நன்றாக ஆராய்ந்து
அறிந்தபின்பு அவற்றை அதற்குரிய செயலில் ஈடுபடுத்தவேண்டும்.

"Never judge someone by the opinion of others"
— Anonymous

11 WATER MANAGEMENT

ஆகாறு அளவிட்டி தாயினும் கேடில்லை
போகாறு அகலாக் கடை.

Though income are limited, no harm
If you don't let expenses exceed them.

பொருள் வரும் வழி சிறியதாக இருந்தாலும் பரவாயில்லை, பொருள் போகும் வழி அளவோடு இருந்தால் எந்த ஒரு வறுமையையும் சமாளித்துக் கொள்ளலாம். வரும் வருமானத்தைத் தக்க வழியில் செலவு செய்து தேவையில்லாத ஆடம்பர செலவுகளைக் குறைத்து சேமிப்பை உயர்த்தினால் நாம் திருப்தியாக வாழ்க்கையை நடத்த முடியும்.

இக்குறள் குடும்பத்தில் நடக்கும் வரவு செலவுகளைப் பற்றிக் கூறினாலும், நாம் பின்பற்ற வேண்டிய நீர் மேலாண்மையையும் ஒப்பிட்டுக்கூறலாம். அதாவது மழையின் அளவு ஏறத்தாழ ஒவ்வொரு வருடமும் ஒரு குறிப்பிட்ட அளவே பெய்கிறது. உலகத்தின் எல்லாப் பகுதிகளிலும் எல்லா வருடங்களிலும் ஒரே அளவானமழை பெய்வது இல்லை. ஒரிடத்தில் ஒரு வருடம் அதிக மழை பொழியலாம், அடுத்த வருடத்தில் அதே இடத்தில் மிகவும் குறைவான மழை பொழியலாம். எனவே அதிக மழை பொழியும் காலங்களில் நீரைச் சரியான முறையில் சேமித்து வைத்தால், அடுத்த ஆண்டு மழை பொழியும் வரை சேமித்து வைத்த நீரை வைத்து, நாம் நமது அனைத்து தேவைகளையும் பூர்த்தி செய்துகொள்ள முடியும்

Everyone should earn money to live a happy life but the amount of earning is not a deciding factor for happy life. The expense which we incur and the savings we achieve are the real deciding factors. So if a person earns more than his actual requirement but spends lavishly and uneconomically then it will lead him into trouble. Ultimately the expense should be within the limit of earnings.

Even though the meaning of this kural specifically describes about the family expense, it can also be appropriately matched with water management.

Let us see how it correlates with the water management system

The overall rainfall across the globe is constant. The total rainfall around the earth is approximately 1000 mm per year. Total quantity of annual rainfall around the globe is 505000 Km^3. In that total rainfall the quantity over the ocean is 398000 Km^3, which becomes saline water and rainfall over the land is 107000 Km^3.

Accordingly, every year the land area around the earth is getting 107000 Km^3 of water as rainfall. Even in this available quantity, a major portion of water will be evaporated and some portion will be percolated into the ground. After all this evaporation, percolation, etc still 36000 Km^3 of water is flowing through rivers. This quantity of freshwater is more than sufficient for human consumption for all their needs including agriculture, industrial and other domestic purposes.

Let us calculate to conclude that this 36000 Km^3 of water is more than sufficient for our needs.

Total run of water is 36000 Km^3 per annum

That is 36000 billion m^3

In terms of litre it is 36000×10^{12} lts

(36000000000000000 lts)

Statistics says that the usage of water towards its purpose is as given below

Agriculture usage 70%

Industrial usage 18%

Domestic usage 12%

Accordingly water available for the domestic use is

$$36000 \times 10^{12} \text{ lts} \times 0.12 \text{ (12\%)}$$

$$= 4320 \times 10^{12} \text{ lts}$$

Now let us see how much water is required per person per day

Current population of the world today is 7.68 billion

So water availability per person per annum is

$$= \frac{4320 \times 10^{12}\,\text{lts}}{7.68 \times 10^{9}\,\text{person}}$$

$$= 562500\ \text{lts}$$

Water availability per person per day is

$$= \frac{562500}{365\ \text{days}}$$

$$= 1541\ \text{lts.}$$

So the nature has been giving 1541 litres of fresh and clean water to each and every person per day which is more than sufficient to meet all our domestic uses like drinking, bathing, washing and sanitation. But we do not save and utilise the runoff water properly. Even 10% of this quantity is enough to meet out our all requirements.

We are creating draught on one side and flood on other side due to improper and inefficient water management. Within the country or even within the state, some areas gets over flooding and other areas suffers of water scarcity and draught.

Let us discuss about better usage of water management.

All the rivers in the country should be interlinked at all possible points. Moreover river paths should not be disturbed by any means such as mining of sand, dumping of waste and encroaching the river bed. Rivers will not search shortest route for attaining their destination, it moves in its own way in order to cover more area and serve more people. We have to allow it to flow on its own path and should not intervene in its path. We can also extend the branches that is tributaries of the rivers to the required place and could also make small check dams to store the water and prevent soil erosion.

Rain water is the product of mega distillation process done by solar energy. The product should not be wasted and completely let out into the ocean. The rainwater after reaching the ground should be allowed to percolate into the ground. We should keep the surface of the earth clean in order to more and more water to infiltrate or percolate deep into the ground. By this, underground water level will be increased. After the saturation of the percolated water into the ground, only the excess water has to be run off over the ground. Nowadays most of the surface area is covered with concrete roads, building and maximum surface area of the ground is dumped with plastic waste and other waste material, which will block the percolation of water into the ground. So we should keep the surface of the earth clean and allow rain water to percolate deep into the ground and hence increase the ground water level.

This runoff water should be collected in ponds, lakes and other water bodies. These water bodies get water during the rainy period and some other water bodies will receive water from the river. Most of the water bodies that were available during ancient times disappeared due to human intervention. All were encroached and converted into residential areas. Due to which most of the rainwater flows into ocean without usage. Now we have to do the following action to enhance ground water level and to meet our water requirements in future.

Water Management:

Water management is the activity of planning, developing, distributing and managing the optimum use of water resources. Here we will discuss some aspects of the water resource management.

Construction of Check Dams:

We should construct check dams across rivers to divert water to local reservoirs, which we can utilize it water for all essential needs.

(Multi purpose river valley projects also help to effectively utilize and conserve water, but during the design and construction phase, it needs to be ensured that the livelihood of people and environment are not getting adversely affected in both upstream and downstream areas)

Rainwater Harvesting:

Harvesting the rainwater from the roof tops and terraces and storing into the reservoirs for reuse. The collected rainwater can also be let out into wells instead of being run off as waste. These activities will raise the groundwater level. Storage of groundwater can be considered as bank balance in our account. Whenever we need we can draw water from the ground and during rainy season we can deposit back into it again.

Drip Irrigation:

Drip irrigation is a type of irrigation which saves water and fertilizer by dripping water slowly to the roots of various crops either into the soil surface or directly onto the root zone through a network of valves, pipes, tubing and emitters. This method saves more water than the traditional watering method.

Reviving of Water Bodies:

In ancient time, there were so many water bodies available globally and each one has a specific purpose. Many of them disappeared during urbanisation and industrialization. Some of them lost their storage capacity without proper maintenance. We have to unearth the disappeared water bodies and put them to use by deepening and cleaning the incoming channels. The plastic waste and other waste which were dumped on the ground and water bodies should be removed to allow proper percolation.

Some of the water bodies which were used by our ancient people are listed here with their Tamil names. Most of them were/had been disappeared now, at least we may satisfy ourselves by recollecting their names.

1. அருவி (water falls)
2. ஆறு (river)
3. சிற்றாறு (jungle stream)
4. நீரூற்று (spring)
5. ஓடை (brook)

6. நீர்ச்சுனை (fountain on the mountains)

7. ஊறணி (A fountain, spring of water)

8. ஊருணி (drinking water tank)

9. இலஞ்சி (A natural or artificial reservoir for drinking (purpose)

10. கிணறு (well)

11. உறைகிணறு (Ring well)

12. நடைக்கிணறு (Large well with steps on one side)

13. கேணி (A square or oblong shaped walled tank)

14. குட்டை (small pool or tank)

15. குளம் (pond)

16. பொய்கை (natural pool)

17. மடு (Deepening area inside the river)

18. குட்டம் (small pond for bathing purposes)

19. தடாகம் (Large pond)

20. கயம் (small tank)

21. ஏரி (lake)

22. கயவாய் (meeting point of river with sea)

23. முகத்துவாரம் (mouth of river over the sea)

24. உப்பங்கழி (Back water salt pan)

25. ஆழிக்கிணறு (well in the sea shore)

These are the ancient water bodies that were identified in Tamilnadu, but these types of the natural water bodies might have been available all over the globe. Each water body has its own specific purpose.

Conclusion:

If all the fresh water bodies which were available during ancient times are revived and brought into usage then there won't be any need of desalination plants and water treatment plants. So as per the great poet's statement, even if we get abundant quantity of anything, it will not be sufficient if we do not utilize it properly and economically.

அளவறிந்து வாழாதான் வாழ்க்கை உளபோல
இல்லாகித் தோன்றாக் கெடும்.

He who lives excessively and lavishly with no limits,
His seeming wealth, will depart, nowhere to be found.

வருமானத்திற்குத் தகுந்தவாறு, பொருளைத் தக்க வழியில் செலவு செய்து, தேவையில்லாமல் பொருள் வீணாவதைத் தவிர்த்து, முறையுடன் வாழாதவனுடைய வாழ்க்கை, இருப்பது போல காட்சி தந்து இல்லாமல் அழிந்துவிடும்.

> "When the well is dry, we know the worth of water"
> — Benjamin Franklin

12 IMPORTANCE OF RENEWABLE ENERGY SOURCE

அறன்ஈனும் இன்பமும் ஈனும் திறனறிந்து
தீதின்றி வந்த பொருள்.

It yields, virtue and joy, the wealth
Acquired capably without causing any harm.

எந்த ஒரு தீங்குமில்லாமல் யாருடைய நலத்தையும் கெடுக்காமல் உழைத்து சம்பாதித்து, நல்ல வழியில் சேர்த்த பொருள், அறத்தையும் கொடுக்கும் இன்பத்தையும் கொடுக்கும்.புதுப்பிக்கக்கூடிய எரிபொருளின் பயன்பாட்டால் சுற்றுப்புறச் சூழ்நிலைக்கு எந்தவிதமான சீர்கேடுகளுமில்லாமல் அதன் ஆற்றலைப் பயன்படுத்தி பூமியில் அனைத்து நாகரிக வளர்ச்சிகளையும் அடைந்துவிட முடியும். ஏராளமான புதுப்பிக்கத்தக்க ஆற்றல்கள் எல்லா நிலைகளிலும் பொதிந்து கிடக்கின்றன. உதாரணமாக நீர்மின்சக்தி, காற்று மின்சக்தி மற்றும் கடல் மின்சக்தி, சூரிய மின்சக்தி, ஆழ்பூமி வெப்ப மின்சக்தி இவைகளின் தன்மைகளை நன்கு ஆராய்ந்து இவற்றின் பயன்பாட்டால் எல்லா வளர்ச்சிகளையும் மனித இனம் அடைய முடியும்.

புதுப்பிக்கத்தக்க எரிபொருளின் பயன்பாட்டால் நமக்கு இன்பமும் கிடைக்கும், இயற்கை வளங்களும் பாதுகாக்கப்படும். இயற்கைவளங்களை பாதுகாப்பதென்பது ஒரு வகையான அறமே! இயற்கைவளங்களை நன்றாக பாதுகாத்து, பின்வரும் சந்ததிகளுக்கு எந்தவிதமான ஊறுபாடும் இன்றி வாழ்க்கை வசதிகளை பெருக்கிக்கொடுக்கவேண்டும் என்பது நம் தலைமுறையின் தலையாய கடமையாகும்.

அப்படியில்லாமல் புதுப்பிக்கமுடியாத கரிம எரிபொருளை, அளவிற்கு அதிகமாக பயன்படுத்தி நாம் நன்றாக வாழ்ந்துவிட்டு வரும் சந்ததியினரை தவிக்கவிட்டு செல்வதென்பது அறம்சாரா அதர்மசெயலாகும். புதுப்பிக்கத்தக்க எரிபொருள் இன்பத்தையும் கொடுக்கும் அறத்தையும் கொடுக்கும். ஆனால் கரிம எரிபொருள் இன்பத்தை மட்டுமே கொடுக்கும் அறத்தை சாராது.

When we earn an asset by our hard work and scrupulously without exploiting anyone wealth will give happiness and honest value to our life. Likewise if we produce power by using good resources like renewable energy by our own hard work without exploiting fossil fuels, it will give happiness by saving our Environment from pollution.

All the energy except tidal energy originate from the sun. Sun is the primary source of all energy. Fossil fuel is termed as ancient solar energy and all other energy are currently forming renewable energy. If we do not utilize the current renewable energy, then all these energy are let out unutilized and merely as waste, that causes the increasing of entropy (பாழாற்றல்) of the universe.

Fossil fuel usage can be termed as someone who sits idly and enjoys his life by using his ancestral property. This type of living is not being considered as virtuous and honest and also there will be no excitement. If we continue to do so then our next generation will struggle even for their food.

Sun is providing plenty of energy to earth. The solar energy spreads everywhere in all the spheres such as atmosphere, hydrosphere lithosphere and biosphere. In the atmosphere it can be seen in the form of wind, in the hydrosphere, the energy is available as water current in rivers, streams and canals. In the lithosphere, ancient solar energy is stored in the form of fossil fuels and in the biosphere the energy is converted and stored as biofuel.

Among all the energy, only fossil fuels are stored like our bank balance, it can be utilized anytime according to our need and requirement. All other forms are dynamic energy, if we don't harness or utilize them on time, then they will be wasted.

Now instead of using the available renewable energy, we are using fossil fuels drastically to fulfill our present power requirements though it is a non renewable source. By utilizing the fossil fuel, it is getting depleted rapidly and at the same time, it is also spoiling the environment by global warming. Utilizing renewable energy source for power production will not have any adverse effect on the environment.

Earth is continuously receiving energy from the sun. That amount of energy is approximately 342 joule per M^2 per second. 70% of energy is returned back to the space by various processes. Remaining 30% of energy is available in the form of hydro, wind, wave, ocean and biofuel. By adapting suitable technology for harvesting the renewable energy efficiently, we can reduce or even stop the usage of fossil fuels.

Encouraging the Renewable Energy Sources:

Renewable energy sources have less carbon intensity. We can reduce carbon dioxide concentration in the atmosphere by encouraging the renewable energy sources.

Carbon Intensity:

Carbon intensity is the quantity of carbon as carbon dioxide emitted to the atmosphere to produce 1 unit (1 Kwhr) of power. This quantity will vary for different energy sources.

Carbon Intensity Data:

Nature of Power	Carbon Intensity
Hydroelectric Power	4
Wind Energy	12
Nuclear Energy	16
Solar Energy	22
Geothermal Energy	45
Natural Gas	469
Coal	1001

We can conclude that least amount of carbon emission will be from the hydroelectric power and most will be from the coal power plant. With the above data we can find out which power source is more environment friendly.

Solar Energy:

Earth is receiving approximately 342 joule per M^2 per second continuously. If we adapt suitable technology, we can meet complete power requirements of the world only through solar energy.

Hydroelectric Power:

Around 23% of solar energy falling on the earth surface is utilized in the hydrological cycle for evaporation.

$$342 \times 23\% = 78.66 \text{ joule/sec/m}^2$$

This much energy is utilized for hydrological cycle per second per square meter.

Total run off water through the river per year is 36000 km^3

$$= 36000 \times 10^9 M^3$$

$$= 36000 \times 10^{12} \text{ kg or lts}$$

Average elevation of the land from the sea level is 800 meters

The potential energy availability (mgh)

$$= 36000 \times 10^{12} \times 800 \times 9.81$$

$$= 28 \times 10^{19} \text{ joule per year}$$

$$(1 \text{ KWHr} = 3.6 \times 10^6 \text{ joule})$$

$$= \frac{28 \times 10^{19}}{3.6 \times 10^6}$$

$$= 7.7 \times 10^{13} \text{ Kwhr per year}$$

The estimated power consumption of total earth is 11×10^{13} Kwhr per year. So maximum energy is available in the water currents along the rivers and streams. The solar energy spent for evaporation in water cycle can be harvested through hydro energy

Wind Energy:

Seven percentage of total energy is utilized for wind formation

$$342 \times 7\%$$
$$23.94 \text{ joule/sec/m2}$$

Moderate high speed wind in the range between 5 metre per second and 25 metre per second can be used for power generation. Low speed wind which is less than 5 metre per second cannot be used as it is less efficient. On the other hand, higher velocity winds with speed of more than 25 metre per second will damage the turbine and the structure.

Wind power generation capacity in India has significantly increased in recent years. As of 31st December 2018, the total installed wind power capacity was 35288 megawatt. India has the fourth largest installed wind Power capacity in the world. Solar radiation used for heating the atmospheric air will be recovered by wind energy

Ocean Wave Energy:

Some portion of the wind energy is transferred to the surface of the ocean that creates the waves in the ocean. Hence Ocean wave energy is also produced by solar energy.

Ocean Thermal Energy Conversion (OTEC):

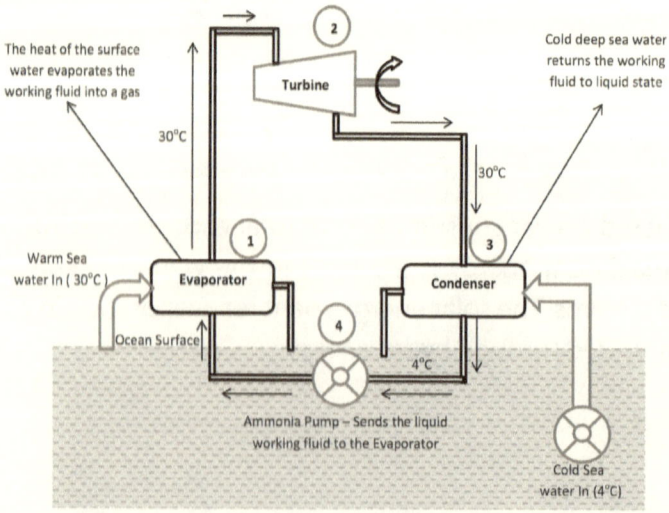

There is a temperature difference between ocean surface water and water at the bottom of the ocean roughly about 200 metres depth. By using this temperature difference, energy can be harvested. This method is called ocean thermal energy conversion method. The temperature difference between the surface water and depth water should be more than 38 °F.

The basic working principle of OTEC is quite simple. The warm water is used to evaporate a working fluid with a low boiling point. The high pressure vapour that is produced drives a turbine-generator to produce electricity. The cold deep seawater is used to condense the working fluid vapour back into a liquid. Finally, a pump drives the working fluid back into the evaporator, to complete the fully closed cycle. In essence, this is similar to how power is produced with a steam cycle found in thermal baseload plants such as coal or nuclear plants, but then in a completely clean and sustainable manner.

Since two third of our Earth is covered by the ocean, huge amount of energy can be obtained from the ocean. This form of energy is also by solar radiation. Sensible heat of water is caused by solar radiation.

Geothermal Energy:

Geothermal energy is stored in interior of the earth. This energy was created during the period of formation of the earth. The energy can be harvested for human uses. This type of energy comes under renewable energy source and is also an ancient solar energy.

Bio Energy:

All bio energy is produced by the process of photosynthesis during the sun light.

Tidal Energy:

The tidal energy also comes under renewable energy source. But this energy is not created by the sun. The moon's revolution around the Earth causes tide formation. During the course of revolution, when the moon is very near to the earth surface, it pulls out the water from the ocean that causes sea levels to rise, causing high tide. When the moon moves away from the earth, its tidal pull gets weaker and water level will recede to its original position, causing low tide. This movement of water from high tide to low tide and vice versa can be harnessed for producing electricity with the help of a turbine. For this technology to work, the difference of water level between high tide and low tide should be more than 5 metre. Energy can not be harvested if it is below five metres.

Here the following data will give power contribution by the each type of power plants in India:

coal (fossil fuel)	79.8%
Hydroelectric	10.1%
Wind energy	4.0%
Solar power	2.0%
Nuclear power	2.9%
Biomass energy	1.2%

Conclusion:

As stated earlier, fossil fuel is termed as ancestral property. From the above data, we can conclude that we are enjoying the life by using our ancestral property. The power production percentage by using fossil fuel is 79.8% which is comparatively huge than the other sources. Now at this juncture, we should utilize higher amount of renewable energy sources like solar, wind, hydro, etc. to avoid pollution.

> *"The nation that leads in renewable energy will be the nation that leads the world"*
>
> — James Cameron

13 FORMATION OF WIND AND WATER CYCLE

இரப்பாரை இல்லாயின் ஈர்ங்கண்மா ஞாலம்
மரப்பாவை சென்று வந்தற்று.

If there are no mendicants, this vast world with people
Will look like wooden puppets that comes and goes.

இரப்பவர் இல்லையானால் இப்பரந்த உலகில் யாசகம் கேட்க யாரும்
இருக்க மாட்டார்கள். உலகம் பொருளாதார நிலையில் சமநிலை
அடைந்து விட்டால் கொடுப்பவர்களும் இருக்கமாட்டார்கள்
வாங்குபவர்களும் இருக்கமாட்டார்கள். அப்படி ஒரு நிலை உலகத்தில்
இருக்கும் பொழுது எந்த ஒரு பொருளும் ஒரிடத்திலிருந்து மற்றொரு
இடத்திற்கோ அல்லது ஒருவரிடமிருந்து வேறு ஒருவருக்கோ
செல்வதற்கு வாய்ப்பில்லை. அந்த வகையிலான உலகத்தின் நிலையை
நினைக்கும் பொழுது மரத்தாலான பொம்மலாட்டம் போன்றே உலக
வாழ்க்கை அமையும். உலக வாழ்வில் உயிரோட்டம் இருக்க
வாய்ப்பில்லை. இதைத்தான் இக்குறளின் மூலம் வள்ளுவர் கூறியுள்ளார்.

In this kural, thiruvalluvar specifically mentioned that if all the people in
the society are equal in all aspects, especially in economic status, then
the world will be looking like a drama acted by puppets. There will be no
merry in that sort of life

For any movement or any material transfer, there should be some
potential difference between the offering and receiving end, otherwise
no material will get transferred from one place to another place or from
one person to another person. If we imagine the world which has no
upper class or lower class people, then all are equal in economic status.
In that condition there will be no need of asking anything from anybody.
No poor will be available for demanding and also no rich will be available
for offering. Economically the society attains equilibrium, in that situation
the world will come to a stand still and there ll be no movement of
material anywhere in the globe.

This Thirukkural concept is appropriately matching with the earth and its surface.

Air and water are the two important things among the five Pancha boodhas. If there is no water and air movement in the globe then the world activities will be like a puppets drama as described by Thiruvalluvar.

If our earth surface is evenly distributed as plains all over the globe, then there will be no air and water movement across the globe. Naturally earth surface is very uneven consisiting of mountains, hills, plains and valleys. This uneven surface pattern with high and low surface areas is the primary cause of creating wind and water movement around the globe and makes the earth more interesting.

How the uneven surface of the earth causes water and wind movement is described here.

Wind Formation:

The earth surface is made up of various patterns of land with both high elevation like hills and low elevation like plains and valleys and also water bodies. Due to solar radiation, there ll be uneven heating due to the uneven surface on the earth which causes variation in temperature in the atmosphere over the globe. This temperature variation creates variation in pressure which causes the movement of air from higher pressure region to low pressure region. This air movement is called wind.

Globally there are two major driving factors of large scale wind pattern. One is the differential heating between the equator and the poles and another one is rotation of the planet. In the equator region solar radiation is very intensive due to less distance between the earth and the sun whereas the intensity of the solar radiation is less in the polar region because of more distance between the earth and sun. The air near the equator is heated up quickly and gets lower in density and rises up.

This creates a void/vaccum in the lower parts of the equator and the air from the polar region will rush towards the equator to compensate the pressure drop. The buoyant air up in the equator will move towards polar regions. In that way air movement occurs continuously between the equator and the polar region and hence the airflow exists all over the globe.

Local Winds:

Local winds are created as a result of various land patterns. Good examples of local winds are sea breeze, land breeze and mountain Valley breeze.

Land Breeze and Sea Breeze:

The differences in specific heat of water and land mass causes the sea breeze and land breeze. The specific heat of the water is more than the specific heat of the land mass (soil). Hence during the day time the land will be heated up quickly than the ocean. So the air near the land will rise up due to the density difference. Then the air from the ocean surface which is relatively cool, will blow towards the land. This is called sea breeze. At night, the roles reverse, the air over the ocean is warmer than the air over the land at night. So the land will get cooled faster than the ocean because of low specific heat and hence the wind blows towards the sea from the land. This is called land breeze.

Mountain and Valley Breeze:

Mountain and Valley breeze are similar to land and sea breeze. At day time the slope of mountains are hot and air from valley blow up to the mountain slope. This is known as Valley breeze. After sunset, the pattern is reversed so that cold air blows from mountain to valley. It is called mountain breeze.

Benefits of the Wind:

Wind movement attempts to balance the unevenness of the atmospheric temperature. Uneven Earth surface causes uneven surface temperature over the globe. No places in the earth will have same temperature at any time. Only the wind movement works continuously to bring down the difference in temperature and pressure in the entire atmosphere over the globe. Uniformity of pressure and temperature can not be possible over the globe at any time, hence air also will not stay standstill. Wind flow also attempts to balance the composition of air uniformly in the entire atmosphere.

The kinetic energy of the Wind can be used directly in many ways. In olden days people used this energy for pumping the water from the well, sailing the boat and in separating grains from the dust during the harvest season in the field. Now wind energy is used for power production.

Wind Power Generation:

Wind energy is the kinetic energy that is associated with the movement of large masses of air. It is estimated that 1% of solar radiation that falls on the earth surface is converted into kinetic energy of the wind. This energy is indirect form of solar energy.

Hydrological Cycle:

Water cycle, also called hydrological cycle, involves the continuous circulation of water between earth surface and atmosphere. In this cycle many processes are involved, the most important are evaporation, transpiration, condensation, precipitation, and runoff. The total amount of water within the cycle remains constant. Sun provides energy for the hydrological cycle.

Solar radiation causes evaporation and transpiration. The water, mainly from the ocean and other water bodies are getting evaporated by solar radiation and the water vapour being lighter than the air rises up. In the upper atmosphere, the water vapour gets condensed, gain mass and comes back to the earth due to gravity as rain.

The rain water, after percolation into the ground starts run off. The run off is due to the difference in elevation between land level and sea level. The average land elevation is 800 meter higher than the sea level and the average ocean depth is 3700 meter lower than the sea level. This huge elevation difference is the main reason for water to run off through the rivers and streams to the ocean. If there are no up and down height difference in the earth surface then there will be no possibility of formation of river, stream and cannel, etc.

Details of Hydrological Cycle

Total quantity of water Evaporation per year: -

Total evaporation and transpiration of water per year	505000 km^3
Evoporation from the ocean	434000 km^3
Evaporation from the land	71000 km^3

Total quantity of Rainfall per year

Total rainfall into the earth	505000 km^3
Rainfall over the oceans	398000 km^3
Rainfall over the land	107000 km^3

Total rainfall over the land per year is	107000 km^3
Total Evaporation on Land per year is	71000 km^3

The Difference between evaporation and precipitation over the land is $107000 - 71000 = 36000$ km^3

(This quantity of 36000 km^3 water is the product or distillate of the natural mega distillation process. All in the Biosphere, from Bacteria to whales and from algae to big trees depend only this water)

Approximately 434000 km^3 of the water is getting evaporated from the oceans but the rainfall over the oceans is only 398000 km^3. On the other hand evaporation from the land is 71000 km^3. But land is getting rainfall of 107000 km^3. Around 36000 km^3 of excess water is received as rainfall on land than its evaporation. This excess rain water falls on the land which was originally owned by the ocean has to return back, hence the rivers crawl towards the oceans.

This run off water from the mountain that flows through the rivers and streams to the ocean provides fresh and pure water for the entire biosphere from micro organisms including plants to humans. The total number of rivers in the world could not be counted. There are about 165 major rivers in the world. These rivers are long and wide enough to be classified as major rivers with large volumes of water flowing through them every day. They have tributaries and provide fresh water to billions of people. There are thousands of smaller rivers, but the exact number of small rivers in the world is difficult to determine. Earth is covered by 773,000 square kilometers of rivers and streams. The total length of the rivers and streams is long enough to circle the globe by 140 times .

The longest rivers in the world are the Nile in Africa and the Amazon in South America. Both rivers flow through many countries. Measuring the length of a river is difficult because it is hard to pinpoint its exact beginning and end. The Amazon is estimated to be between 6,259 kilometers and 6,800 kilometers long. The Nile is estimated to be between 5,499 kilometers and 6,690 kilometers long. The Amazon carries more water than any other river on Earth. Approximately one-fifth of the freshwater entering the oceans comes from the Amazon.

Hydroelectric Power:

The run off water from the rivers not only provides doemestic, Industrial and and agricultural usage, it also can be used to produce energy. The energy of flowing river water comes from the force of gravity, which pulls the water downward. The steeper the slope of a river, the faster the river moves and the more energy it has. The movement of water in a river is called a current. The current is usually strongest near the river's source,since the slope is also very high.

A dam is a barrier that stops or diverts the flow of water along a river. Humans have built dams across the river some thousands of years ago. In 1882, the world's first hydroelectric power plant was built on the Fox River in the U.S. Now thousands of hydroelectric plants have been built on rivers all over the world. About 19 percent of power in the world, comes from hydroelectric plants. China is the world's largest producer of hydroelectric power.

Hydroelectric power is renewable because water is constantly replenished through precipitation and hydroelectric plants do not burn fossil fuels, they do not emit pollution or greenhouse gases.

Now we have to recall Thirukkural again

இரப்பாரை இல்லாயின் ஈர்ங்கண்மா ஞாலம்
மரப்பாவை சென்று வந்தற்று

Here the ocean is being continuously receiving water from the land due to the huge elevation difference between the land and the ocean. The average ocean depth from the average land level is (800 m + 3700 m) 4500 meters. (The average land elevation is 800 meter more than the sea level and the average ocean depth is 3700 meters lower than the sea level). If this much elevation difference is not available between the ocean and the land, then there will be no movement of water from the land into the ocean. In that condition there will be no chance of formation of rivers and streams. Without rivers and streams, the world activities becomes zero. Here ocean is the receiver and mountains are the donor. Water is being circulated between them and make the earth lively and beautiful.

Wind formation also plays a major role for making the globe as cheerful as it is presently.

ஈவார்கண் என்னுண்டாம் தோற்றம் இரந்துகோள்
மேவார் இலாஅக் கடை.

Ah, what is there to brag about those who give
If there be none to seek and happily receive.

பொருள் இல்லை என்று பெற்றுக்கொள்ள ஒருவனும் இல்லை எனில் பொருள் வைத்திருப்பவனின் புகழ் எவ்வாறு வெளிப்படும், கொடுப்பவனும் வாங்குபவனும் இருந்தால்தான் இந்த உலகத்தில் அனைத்து நிகழ்வுகளும் நடைபெறும், அப்போதுதான் உலக இயக்கம் உயிரோட்டம் உள்ளதாக இருக்கும், நீர் சுழற்சியை எடுத்துக்கொண்டால் கடல் நீரை உள்வாங்குகிறது மலையும் மேடான நிலப்பரப்பும் தண்ணீரை கடலுக்கு அனுப்பிக் கொண்டிருக்கிறது, இந்த நீரோட்டமே உயிரின் ஆதாரம். கடல் மட்டம் கீழே இருப்பதனால்தான் ஆறுகளும் நீரோடைகளும் உயிரின் ஆதாரமாக இயங்குகின்றன.

How does a person's charitable reputation be manifested if there's no one to receive it, all the events in this world take place only if there's a giver and a taker. That will be the lifeline in which the world functions.

Take for instance water cycle, water enters seas and oceans, from mountains and higher landscapes through rivers and water bodies which serve as a source of life because sea level is lower than land level.

> *"If the earth is a mother then rivers are her veins."*
>
> — Amit Kalantri

14 ENERGY SHARING BETWEEN SUN AND THE EARTH

வகுத்தான் வகுத்த வகையல்லால் கோடி
தொகுத்தார்க்கும் துய்த்தல் அரிது

One may accumulate wealth worth crores, but can consume
Only as ordained by the ordainer.

மனிதன் எவ்வளவுதான் சொத்து சுகங்களைச் சேர்த்து வைத்தாலும் அவனால் அனைத்தையும் அனுபவிக்க முடியாது. ஒரு குறிப்பிட்ட அளவே அவனால் நுகர முடியும். அவன் சேர்த்து வைத்த சொத்துக்கள் யாவும் மற்றவர்களுக்குப் பயன்படுமாறு அமைத்துக்கொண்டால் அவன் வாழ்வும் சிறக்கும். அவனால் மற்றவர்கள் வாழ்வும் சிறந்து ஓங்கும். நமது சூரியகுடும்பத்திலேயும், சூரியன் உற்பத்தி செய்யும் மொத்த ஆற்றலையும், பூமியோ அல்லது பிற கோள்களோ பெற்றுக்கொள்வது இல்லை. சூரியன் உற்பத்தி செய்யும் மொத்த ஆற்றலில் சுமார் 0.1 சதவீதத்திற்கும் குறைவாகவேஎல்லாகோள்களும் பெற்றுக்கொள்கின்றன.

பூமியும் அதுபோலவே சூரியனிடமிருந்து பெறும் அவ்வளவு ஆற்றலையும் பயன்படுத்துவது இல்லை, மிகக் குறைந்த அளவே தாவரங்களில் ஒளிச்சேர்க்கைக்காக பயன்படுத்துகிறது, சூரியனிடமிருந்து பெற்ற அனைத்து ஆற்றல்களையும், நீர்சுழற்சிக்காகவும் மற்றும் காற்று மண்டல சுழற்சிக்காகவும் பயன்படுத்தி பூமியின் வெப்பநிலையைச் சீர்படுத்திக்கொண்டிருக்கிறது. அவ்வளவு வெப்பத்தையும் பூமி தனக்காக உட்கிரகித்துக்கொண்டால் பூமியின் வெப்பம் மிக வேகமாக அதிகரித்து பூமி பல நூற்றாண்டுகளுக்கு முன்பே எரிந்து சாம்பலாகியிருக்கும்.

The wealth could not be utilised completely by an individual even though he may possess the same abundantly. He could use only a portion of it. If a person understands this concept and starts distributing his excess wealth to others who are in real need, his life will become more peaceful and meaningful.

We could correlate this with Sun's energy production and distribution.

The Sun produces very huge amount of energy per second and scatters to all the planets including the earth. But Earth is getting only one out of 2 billionth of sun's total energy. Though it receives only a small percentage of the total energy, it is not consuming all the energy completely, it is contributing to other activities which are essential for life existence in Biosphere.

Let us discuss here about our solar system and it's energy distribution.

Sun

Sun is the star, located in the centre of solar system, mother of all planets. It is nearly a perfect sphere of hot plasma, its diameter is 1.39 million kilometres. About 99.86 percentage of total mass of the solar system is occupied only by the sun. Remaining 0.14% of the mass is occupied by all other planets including the earth. The total volume of the sun is 1.4×10^{27} cubic meters. Just for imagination, about 1.3 million earths could fit inside the sun. The mass of the sun is 1.989×10^{30} kgs. That is about 333000 times the mass of the earth. The distance between the center of the sun and the surface of the Earth is 150 million kilometers. It is called as one astronomical units (AU).

The constituents of the sun:

Hydrogen	73.0%
Helium	25.0%
Oxygen, carbon, iron & neon	2.0%

Sun is producing energy by the fusion reaction of hydrogen into helium. It currently fuses about 600 million tons of hydrogen into Helium every second .596 million tons of helium is produced by the hydrogen fusion reaction. Remaining 4.0 million tons are mass defect. This 4 million tons of mass is being converted into energy every second.

Total energy produced by the sun per second

$$E = mc^2$$

Mass $= 4.0 \times 10^6$ ton

$= 4 \times 10^9$ kgs.

Velocity of light	$= 3.0 \times 10^8$ m/sec.
Energy	$= 4.0 \times 10^9 \ (3.0 \times 10^8 \ \text{m/sec})^2$
Energy produced	$= 36 \times 10^{25}$ joule/sec.

Total energy production by the sun per second is 36×10^{25} joules.

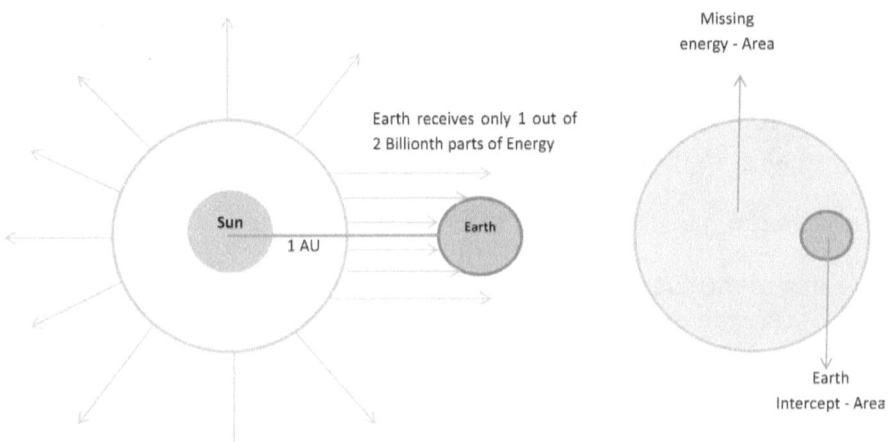

Total volume of imaginary sphere is 1.414×10^{25} km³

The total surface area of the imaginary sphere is 2857.14×10^{25} km²

The Area of intercept by the Earth on Total solar radiation is 127.52×10^6 km²

Let us have some calculations to conclude fraction of the incoming energy of the earth to the total solar energy produced by sun per second.

Imagine a large sphere which has the radius of one astronomical unit, that is the distance between the centre of the sun and the surface of the earth.

1 astronomical units (R)	$= 150 \times 10^6$ k.m
Volume of the sphere	$= 4/3 \ \pi \ R^3$
	$= 4/3 \ \pi \ (150 \times 10^6)^3$ km³
	$= 1.414 \times 10^{25}$ Km³

This is an imaginary sphere, just calculated to compare the total area of solar radiation that spreads in all direction near the earth surface in which the area of earth intercepts. The ratio of area intercepted by the earth to the imaginary sphere area could be only one out of two billionth parts.

Overall surface area of the imaginary sphere $= 4 \times \pi R^2$

$$= 4 \times \pi (150 \times 10^6)^2 \text{ square k.m}$$

$$= 282857.14 \times 10^{12} \text{ square k.m}$$

So the available area of the solar radiation in all directions is as huge as 2.828 × 10^{17} square kilometers. But the area of earth intercepting it is very small as derived below

The radius of the earth (r) is 6370 k.m

The area of the circle equal to the earth radius $= \pi r^2$

$$= \frac{22 \times 6370 \times 6370}{7}$$

$$= 127.52 \times 10^6 \text{ km}^2$$

But the entire vicinity area of radiation

$$= 282857.14 \times 10^{12} \text{ km}^2$$

So the fraction of earth interception area in the total covered radiation zone

$$= \frac{127.52 \times 10^6 \text{ km}^2}{282857.14 \times 10^{12} \text{ km}^2}$$

$$= \frac{0.450}{10^9}$$

$$\text{Approximately} = \frac{1.0}{2 \times 10^9}$$

(That is one out of two billionth of the total solar radiation area)

We could also arrive the samevalue by another method by calculating the ratio of incoming solar energy received by the earth to the total energy produced by the sun.

Let us see the calculation

Total energy produced by the sun per second

36×10^{25} joule/sec.

Earth input radiation from the Sun = 342 joule per second per square metre

(Earth is receiving 342 joule per second per square meter)

Earth total surface = 5.1×10^{14} square metre

Total energy received by the earth per second

$$= 5.1 \times 10^{14} \times 342$$

$$= 1.74675 \times 10^{17} \text{ joules per second}$$

Approximately

$$= 1.8 \times 10^{17} \text{ joules}$$

Therefore the fraction of Earth receiving energy to the total energy produced by sun

$$= \frac{1.8 \times 10^{17} \text{ joules second}}{36 \times 10^{25} \text{ joule/sec.}}$$

$$= \frac{1.8 \times 10^{17} \text{ joules second}}{3.6 \times 10^{26} \text{ joule/sec.}}$$

$$= \frac{1.0}{2 \times 10^{9}}$$

(One out of two billionth parts)

The Earth is not keeping all the energy that it receives from the sun. The earth utilizes the energy for water and air circulation. Some small portion, that is only 0.02% of the total energy that falls on the earth surface is taken by the plants for photosynthesis. The entire biosphere from algae to big trees and from bacteria to whales including humans are depending on this 0.02% of solar energy utilized for plant photosynthesis.

தாளாற்றித் தந்த பொருளெல்லாம் தக்கார்க்கு
வேளாண்மை செய்தற் பொருட்டு

The purpose of someone's hard earned money is
To be deployed to help the deserving.

அறவழியில் சேர்த்த அத்தனை செல்வங்களும் தகுதியானவர்களுக்குக்
கொடுத்து உதவுவதற்காக மட்டுமே. தாம் மட்டுமே நுகர வேண்டும்
என்று நினைத்தால் விரைவில் செல்வம் முழுவதும் அழிந்து நிற்கதி
ஆகிவிடுவோம்.

> "Every 24 hours, enough sunlight touches the earth
> to provide the energy for the entire planet for 24 years"
>
> — Martha Maeda

CREATURES IN THE ECO SYSTEM – ROLE VS APPEARANCE

உருவுகண்டு எள்ளாமை வேண்டும் உருட்பெரும்தேருக்கு
அச்சாணி அன்னார் உடைத்து.

Deride not a person seeing their appearance
There are many persons who are like a small lynchpin that
make a large chariot roll.

பூமியின் உயிர்க்கோலத்திலும் சில உயிர்கள் தோற்றப்பொலிவுடனும்,
சில உயிர்கள் அவலட்சணமாகவும் இருக்கும். நாம் அவைகளின்
உருவத்தை வைத்து மட்டுமே அவைகளின் முக்கியத்துவத்தை அறிய
முற்படக்கூடாது. அதனுடைய பங்களிப்பு இந்த உயிர்க்கோலத்தில்
என்ன? என்று ஆராய்ந்து அதன்படி அவ்வுயிரின் தன்மையை
அளவிடவேண்டும்.

Some people may look unpleasant. They may have a dark complexion, they don't even attain minimum height, their face may look ugly but we should not underestimate and dishonor them based on their external appearance. They may be very important persons similar to a Lynchpin which is made up of iron and used as a guard for rotating wheel.

If the Lynchpin is not available on the outer side of the axle of the rotating wheel, then the wheel will come out from the axle and finally the cart will collapse. The Lynchpin does not look nice – it'll have a ragged appearance with a dark and oily surface. But its importance in a moving cart cannot be underestimated. Without the lynchpin, we cannot drive the cart. Axle may be the prime moving part of the cart. But this Lynchpin is more important than the axle.

Similarly, in our ecosystem, some creatures may look very unpleasant, it may not have fancy colours and charming look but its contribution towards society and environment would be commendable.

Let us discuss about some of the creatures which might be ugly in appearance but are excellent in their performance.

Role of Animals in Ecosystem:

Many creatures live together in the biosphere causing biodiversity. Each and every creature has its own value. Each creature is doing some specified work in the environment by their routine work as part of their existence. All the organisms are interlinked with each other in the food chain with others. Every animal in the ecosystem plays a vital role in the wellbeing of the planet. If one species is extinct due to some disturbance in the environment, it can have a significant effect throughout the rest of the ecosystem. Even a small earthworm is a crucial worker in the society of the nature.

Earthworm:

Earthworms may lack the charming looks. But their contribution to the world is significant. This lowly creature plays a vital role in the natural soil ecosystem. They are also called as "ecosystem engineers" because they significantly modify the physical, chemical and biological properties of the soil profile. This modification can influence the habitat and activities of other organisms within the soil ecosystem for further improvement of the soil structure.

Earthworms are known as the friends of farmers due to the following reasons. They improve the fertility of the soil in different ways and therefore they are most important for agriculture. Actually the burrowing and soil feeding habit of earthworms make the soil porous which permit both aeration and quick absorption of water. It also permits easy penetration of plant roots. Their burrowing habit can increase the water infiltration rate up to 10 times more than the normal. Decomposition of dung and plant litter by the earth worms makes 2 to 20 tons of organic matter per hectare each year.

Earthworms also bring fresh subsoil to the surface which is still finer and rich in organic matter. It is estimated that the number of earthworms in an acre of land reach up to more than 25,00,000 which can bring more than 18 tons of deep sub soil to the outer surface area in one year. The faeces of earthworm contains nitrate, calcium, magnesium, potassium and phosphorus which are very much essential for plant growth.

Earthworms also control the alkalinity and acidity of the soil to provide better condition for plant growth. Decomposition of earthworms after their death will increase the organic constituents of the soil. Thus the earthworms make the soil fertile to a great extent. Earthworms are also called as a natural ploughmen.

Ayurvedic and Unani system of therapy suggest that this worms were used in making medicine for the cure of variety of diseases such as bladder stone, jaundice, pile and rheumatism etc. Even today these are used for making various medicines in India as well as in other countries.

Caterpillar:

Caterpillar also does not have an elegant appearance. Many people will get annoyed and will be getting irritated on its touch. The Caterpillar is the second stage in the butterfly life span, the first is being larva stage.

Caterpillar stage is the most important stage in the life cycle of the butterfly. Only in this stage, lot of activities are happening inside to build up adult butterfly. Butterfly is very charming and cheerful coloured insect but the previous stage of butterfly is the Caterpillar stage, which is ugly and unpleasant.

If all Caterpillars are extinct from the earth due to any of the human activities such as global warming or acidification or deliberately destroyed by human for its unpleasant looks, then butterflies will not be available for pollination. This will lead to lot of food scarcity and will push the entire world in to severe famine. Butterflies are also vital indicators of healthy environment and ecosystem. Butterflies are diverse group of insects consisiting approximately 20,000 different species in all over the globe.

Pollination:

Nearly 70% of the food crops depend on pollinators like butterfly and honey bee for pollination. Therefore their existence is very important for food production. Some species are also doing pest control. For example, the harvester butterflies eat aphids while they are in Caterpillar stage.

Butterflies are particularly more sensitive to climate change. Scientists monitor the existence of butterflies as a method to monitor the environment and ecosystem. The area which is having more population of butterflies is considered as a sign of healthy environment. So we should protect them from its extinction. Protecting the butterflies will improve the environment worldwide and enrich the lives of people in the present and also in future.

Frogs:

Frogs are not given much attention due to their appearance, but they are more important to man than its use for high school dissection experiments. Frogs act as a bioindicator,which means they indicate the health of the environment. Frogs have the ability to live in land as well as water.

Since frogs are most often be the first animal to react the biological and chemical hazards, they are helpful to warn humans to take action against pollution. The skin of the frog is very porous and permeable which allows substances present in the environment to be absorbed with in the fatty tissue. By analysing it's skin we can find out the nature of contamination. In this way, it indicates environmental pollution of both land and water bodies.

Frogs also help in insect control. They are also food for many animals like snake, ducks etc. So they are very much needed for maintaining balance of ecosystem

Conclusion:

"Don't judge a book by its cover"

Earth worms, bees, frogs and caterpillars do not look beautiful. They look ugly and unpleasant. But their importance in the environment cannot be undervalued. They only provide food for entire biosphere.

They play a vital role in the earth's food cycle, The pollinators are extremely important to the world food Security.

கணைகொடிது யாழ்கோடு செய்வித்தாங் கன்ன
வினைபடு பாலாற் கொளல்

Cruel is the arrow which is straight, the crooked lute is sweet,
Judge by their deeds the many forms of men you meet.

கூர்மையான அழகிய வடிவமைப்பை உடைய அம்பு கொடுமையை உண்டாக்கும். அதே நேரத்தில் வளைந்தும் நெளிந்தும் கோணலான வடிவமைப்பு உடைய யாழ் அருமையான இசையைத் தரும். எனவே மனிதர்களையும், மற்ற மிருகங்களையும் அதன் உருவத்தை மட்டும் வைத்து நன்மைகளையும், தீமைகளையும் தீர்மானிக்கக் கூடாது. அவைகளின் செயல்பாட்டை வைத்தே எல்லாவற்றையும் தீர்மானிக்கவேண்டும்

A beautiful and sharp arrow makes dreadful happenings but a "violin" which has uneven shape and hard looks gives fabulous music. Hence we should not praise or underestimate a person or animal or any creature by it appearance and we should conclude only based on their performance, uses and good behavior.

> *"The caterpillar does all the work
> but the butterfly gets all the publicity."*
>
> — George Carlin

16 OZONE DEPLETION

ஆய்ந்தாய்ந்து கொள்ளாதான் கேண்மை கடைமுறை
தான்சாம் துயரம் தரும்.

Assess and re assess before taking someone as a friend,
Else the resulting anguish will be fatal.

ஒருவரது நல்ல குணங்கள் தீய குணங்கள் இவற்றை ஒருமுறைக்குப்
பலமுறை நன்றாக ஆராய்ந்து உறுதியான முடிவு எடுத்தபின் அவரிடம்
நட்பு கொள்ள வேண்டும். ஆராயாமல் கொள்ளும் நட்பு மிகுந்த
துன்பத்தைத் தரும். சில சமயங்களில் நம் உயிருக்கே கூட ஆபத்தாக
முடியும்.

We should analyze the character and quality of a person completely
before having friendship with him. If required, we can review many
times before taking the final decision of having friendship with him.
Otherwise we will get into big trouble by selecting a unworthy person
as our friend. Sometimes his friendship may even lead to threatening
of life.

Similarly when we develop any new technology, we must analyze
the impact of that technology on the environment in future. If we do not
take proper attention to its effects, then we will be suffering a lot in
future.

Let us discuss the above with an example of invention of refrigerators.

Since beginning of the 20th century, global warming started due
to the accumulation of greenhouse gases. Scientists developed
refrigerators for domestic use as well as for industrial purposes and
people also started using refrigerator for their comfort and to avoid
food waste. The working fluid inside refrigerator was chosen as freon
(CFC, chlorofluorocarbon)

Structure of one of the chlorofluorocarbon:

$$
\begin{array}{ccc}
 & Cl & F \\
 & | & | \\
Cl - & C - C & - F \\
 & | & | \\
 & Cl & F
\end{array}
$$

Positive and Negative Effects of Freon:

CFC (chlorofluorocarbon) consists of carbon, chlorine and fluorine atoms. CFC has very good qualities as a working fluid in refrigerators, but it causes many negative impacts to the environment.

Positive Aspects of Chlorofluorocarbon as a Refrigerant:

These compounds are highly stable, non flammable, water insoluble, tasteless and odourless. The important property of a refrigerant is their (volatility) boiling point close to 0°C. These physical and chemical properties make them ideal for use as a refrigerant in the air conditioners, freezers and refrigerators.

Negative Impacts of Chlorofluorocarbon:

Chlorine atom made by decomposition of CFC produces harmful effects to the environment.

Ozone in the stratosphere region is protecting us as an umbrella from the harmful ultraviolet radiation. But this chlorine atom causes ozone depletion in the stratosphere region. Most of the chemicals, when released into the atmosphere will get rapidly broken down into smaller harmless compounds by the chemical reaction in the lower atmosphere but this CFC is so stable and unreactive that they survive to reach the highest level of the stratosphere region and spreads all over the world. At this high altitude the intensity of ultraviolet radiation is so high that even the stable CFC molecule splits and releases chlorine atom.

$$CF_3Cl_3 \xrightarrow{\text{(UV rays)}} CF_3Cl_2 + Cl\cdot$$

This released chlorine atom will stay in the stratosphere region for more than 200 years and destroy more than 1 lakh ozone molecule in the following manner.

Stratosphere Region and Activities of Ozone Layer (12–50 km):

Ozone molecules in this layer absorb high-energy ultraviolet (UV) rays from the Sun, converting the UV energy into heat. More than 90% of the ozone in the atmosphere are staying only in the stratosphere region. In this region, the ozone – oxygen cycle happens continuously and destroy harmful UV rays. Ozone concentration also is maintaining steadily by this ozone – oxygen cycle.

Ozone–Oxygen Cycle:

An oxygen molecule is split by stronger frequency UV rays into two separate oxygen atom

$$O_2 \xrightarrow{\text{(UV rays)}} 2\,(O)$$

Each oxygen atom then quickly combines with the oxygen molecule and forms ozone molecule

$$(O) + O_2 \xrightarrow{\text{(UV rays)}} O_3$$

The ozone molecule formed by the reaction will absorb UV rays, those having wave length between 240 and 310 nano meters. During the absorption of the UV rays the ozone molecule is split into oxygen molecule and the atomic oxygen.

$$O_3. \xrightarrow{\text{(UV rays 240–310 nm)}} O_2 + (O)$$

This atomic oxygen quickly reacts with the another oxygen molecule and reforms ozone.

$$(O) + O_2 \xrightarrow{\text{(UV rays)}} O_3$$

The overall effect is conversion of UV rays into heat without any net loss of ozone. This cycle keeps the Ozone layer in a stable and constant thickness. and the harmful UV radiation is being filtered by the oxygen–ozone cycle.

When the chlorofluorocarbon reaches stratosphere region, it gets split by the energetic UV rays and releases the chlorine atom. This Chlorine atom intervenes in the oxygen–ozone cycle and destroys Ozone molecules.

$$CF_3Cl_3 \xrightarrow{\text{(UV rays)}} CF_3Cl_2 + Cl\bullet$$

The freed Chlorine atom reacts with ozone molecule and destroys it.

$$Cl\bullet + O_3 \longrightarrow O_2 + ClO$$

This ClO reacts with another ozone molecule and releases Chlorine atom again.

$$ClO + O_3 \longrightarrow 2O_2 + Cl\bullet$$

The atomic oxygen releases during the ozone – oxygen cycle reacts with ClO and releases Chlorine atom.

$$ClO + (O), \longrightarrow O_2 + Cl\bullet$$

It is estimated that one atom of chlorine can stay more than 200 years in the stratosphere region and destroys as many as 100000 Ozone molecules. As the ozone layer becomes thinner, the protective filter provided by the atmosphere is progressively reduced. Consequently human beings and the environment are exposed to higher UV radiation levels. Higher UV levels have the greatest impact on the health of humans, animals, marine organism and plants. Ozone concentration should be maintained at 350 Dobson unit (DU) in the stratosphere region.

The Dobson Unit is the most common unit for measuring ozone concentration. One Dobson Unit is 0.01 millimeters thick. In stratosphere region, ozone concentration is 350 Dobson units. It means if all the ozone molecules are compressed around the globe at sea level at 0°C and one atmospheric pressure, then the ozone layer will be 3.5 mm thick. During the beginning of summer the Antarctica ozone concentration will reach less than 200 DU. The area with less concentration of ozone above the Antarctica is called as ozone hole. The hole is developed due to the ingress of CFC in the stratosphere region.

Kinds of UV Radiation:

UV–A radiation 315–400 nm

UV–B radiation 280–315 nm

UV–C radiation 100–280 nm

Wavelength of UV–C is very minimum hence it possess high potential danger than UV–B and UV–A.

When sunlight passes through the atmosphere, all UV–C rays are completely filtered by the Ozone layer in the stratosphere region and also approximately 90% of the UV–B radiation is filtered in the same region by ozone. Therefore the UV radiation that is reaching the earth surface is largely composed of UV–A with a small amount of UV–B component. UV–C is completely filtered in the stratosphere region itself.

Chlorofluorocarbon as a Greenhouse Gas:

Chlorofluorocarbon is not only destroying zone layer in the stratosphere, but also it act as highly potent greenhouse gas. It's global warming potential is higher than the CO_2 and methane. Global warming potential is the main parameter which determines the intensity of the greenhouse effect.

Some of the greenhouse gases and their global warming potential are given below:

Greenhouse Gases	GWP (Global Warming Potential)
CO_2	1.0
CH_4	62.0
N_2O	275.0
CFC–12	7900.0

CFC–12 has global warming potential value as 7900. It means 1.0 kg of CFC–12 emitted into the atmosphere is equal to 7900 kgs. of Co_2 releasing into atmosphere. This is such a high potential harmful greenhouse gas. Not only CFC 12, all CFC's have more or less the same GWP value.

Conclusion:

The refrigerant CFC (Chlorofluorocarbon) was developed for the usage in the refrigerators as a working fluid. As a refrigerant it has performed well in the refrigerators and continue to provide cooling effect. But its effects on the environment is very much harmful. It is not only destroying ozone layer in the stratosphere, it also act as greenhouse gas and leads to the global warming.

However Hydroflurochlorcarbon (HCFC) and Hydrofluorocarbon (HFC) were chosen as the better replacement of CFC. They were introduced in 1980 since their ozone depletion potential is zero. The main advantages of HCFCs and HFCs over CFCs are they are less stable and more reactive with their additional hydrogen atom(s). Hence they could easily get break down in the troposphere before reaching the stratosphere. Their atmospheric life time is usually less than 14 years whereas the life time of CFC is more than 200 years.

If the scientists were analysing properly as mentioned in the kural they would have replaced HFC in place of CFC which might not have been depleted the ozone layer to such an extent.

நாடாது நட்டலிற் கேடில்லை நட்டபின்
வீடில்லை நட்பாள் பவர்க்கு.

Nothing can be more destructive than
Accepting a friend without assessing.

ஆராயாமல் கொண்ட தீய நட்பைவிட மோசமானது உலகத்தில் எதுவுமே இல்லை. அப்படி ஒருவர் தீய நபரிடம் நட்பு கொண்டுவிட்டால் அவருக்கு அதிலிருந்து விடுதலையே கிடையாது.

The friendship without analysing the character of the person will give more problem to us. It is very difficult to leave the friendship when we find any immoral character on them.

> "Some of the most poisonous people come disguised as friends and family"

17 SEVERITY OF SEA LEVEL RISE

வருமுன்னர்க் காவாதான் வாழ்க்கை எரிமுன்னர்
வைத்தூறு போலக் கெடும்.

His life who doesn't guard against the forthcoming evil,
Like straw before the fire shall swiftly be destroyed away

ஏதேனும் ஒரு தீங்கு வரும்முன் அனுமானித்து, அதன் தன்மையை
உணர்ந்து அது பிற்காலத்தில் நமக்கு கொடுக்கும் தீங்கினை ஆராய்ந்து
அதை தடுக்காதவனுடைய வாழ்க்கை அல்லது செய்யத் தவறிய
மன்னனுடைய நாடு, சமுதாயம், தீயின் முன் உள்ள வைக்கோல் படப்பு
போல கருகி எரிந்து சாம்பலாகிவிடும்.

One should foresee the possibility of any troubles that might happen in future and take precautions to avoid such happenings. Always "Prevention is better than cure". If a person or the king of a country does not foresee and prevent such things, then their life will disappear like paddy straw which is kept in front of the fire.

One of the disastrous effects of global warming is the rising of sea level. We should take corrective action immediately to stop global warming, Otherwise the entire land area will get submerged under water. Future generations will suffer a lot like Tsunami victims. In such conditions only aquatic organisms will exist on the earth and no terrestrial life will exist.

As we know the land area occupies only one third of the total area of the earth and the remaining area is covered by water. Earth is the only planet in the solar system where life is existing and if we allow it to get submerged under water then where do we live?

Water exists throughout the Globe in all the three states viz solid, liquid and vapour. In glaciers and polar regions it exists in solid state, in the atmosphere it is in vapour state and in the water bodies it is exists in liquid state.

The quantity of water existing in each state should not be disturbed for maintaining global balance. There is a dynamic equilibrium among all three states in relation with the global temperature. If the global temperature increases then the quantity of water in solid state (ice sheets) will reduce due to melting and flow into the ocean and at the same time, water vapour (moisture) concentration in the atmosphere will increase due to more vaporisation. Reduction in the quantity of solid state and increase in the quantity of vapour state will further accelerate global warming. Reduction of the polar ice sheet area will also reduce the albedo effect and subsequent increase in moisture concentration in the atmosphere will trigger greenhouse effect. Both are not advisable.

Water Distribution in the Earth

The total volume of water available on Earth is estimated as 1.386 billion km^3. Out of which 97.5% is saline water and only 2.5% is the fresh water available by all means. Out of the 2.5% of the fresh water, almost 1.74% of water is locked up in the glaciers and in ice caps. Maximum part of the balance available freshwater is lying underground as ground water. Only less than 0.5% of freshwater is on the surface which are available in rivers, lakes, ponds and streams. Major portion of fresh water is locked up in the glaciers and ice caps as solid (ice).

Sea Level Rise:

It has been estimated that the sea level rose by 2.7 millimeters per year from 1993 to 2004. And it has been increased to 3.5 millimeters per year in 2019. It is estimated that, each degree Celsius of temperature rise triggers a sea level rise of approximately 2.3 meters. Meteorologists expect that the rate will further increase during the 21st century. This acceleration mostly is due to human caused global warming.

THE POSSIBLE CAUSES OF SEA LEVEL RISE

1. Thermal expansion of seawater
2. The melting of ice sheets at the glaciers
3. Over mining of underground water

1. Thermal Expansion of Seawater

Thermal expansion of the ocean water contributes 42% of sea level rise due to global warming. Density of water is maximum at 4°C. and above this value the density will decrease. Due to the rise in temperature, density of water decreases and thereby its volume increases considerably. Thus when the overall temperature of the water in the ocean increases, it naturally expands and hence will increase the ocean level.

2. Melting of Polar Ice Sheets and Glaciers

Melting of ice from ice caps and ice sheets contribute to 44% of the total sea level rise. The contribution of temperate glaciers – 21%, Greenland glaciers – 15% and Antarctica glaciers and ice sheets – 8%.

Temperate Glaciers:

A temperate glacier (warm glacier) is a glacier that's essentially at the melting point, so solid and liquid states of water coexist with glacier ice. A small change in temperature can have a major impact on melting of temperate glacier. Temperate glaciers exist in the continents of North America, South America, Europe, Africa, Asia and on both islands of New Zealand. When compared to Atlantic ice sheet and Green land ice sheet, the volume of temperate glacier is very low. Despite this, the contribution of sea level rise due to temperate glacier is high because of its warmness. Slight temperature raise will trigger melting of ice in temperate glaciers.

Antarctic Ice Sheet:

The Antarctic Ice Sheet is the largest store of frozen freshwater. The land area in south pole of the earth, Antarctica, is covered by the larger ice sheet than all. It covers an area of about 14.6 million km^2 and contains between 25 and 30 million km^3 of ice. Around 70% of the fresh water on the Earth is contained in this ice sheet. About 90 percent of the world's ice is present in Antarctica. The average height of the sheets in Antarctica is 2,133 meters. If all of the Antarctic ice melted, then the sea level around the world would rise by 61 meters. But the average temperature in Antarctica is –37°C, and so there is little or no chance of melting easily. In fact, in most parts of the continent, the temperature never gets above freezing point. So the possibility of Antarctica ice sheet melting is very remote.

The Greenland Ice Sheet:

This is a vast body of ice covering 1,710,000 square kilometres, roughly 80% of the surface of Greenland. It is the second largest ice body in the world, after the Antarctic ice sheet. Its thickness is generally more than 2 km and over 3 km at its thickest point. Total volume of ice is approximately 2.85×10^6 k.m³. If the entire 2,850,000 cubic kilometres of ice melted, it would lead to a global sea level rise of 7.2 m. Greenland is closer to the equator than Antarctica, so the temperature is more than that of Antarctica and so the ice is more likely to melt quickly than Antarctica ice.

Artic Ice Sheet:

At the other end of the world, the North Pole, there is a ocean (Arctic ocean) covered by ice. This ice is not so thick as at the South Pole. The ice floats on the Arctic Ocean and therefore, if it melts sea levels would not be affected. The water locked up in the temperate glaciers and Greenland ice sheets are only the more vulnerable areas which will get affected even by small increase in temperature in the earth surface.

Estimated Sea Level Rise

Let us calculate the sea level rise if all the ice melts in the polar region

Antarctic Ice Sheet:

Area covered	$= 12.1 \times 10^6$ km²
Average thickness	$= 2.396$ km
Quantity of ice	$= 29 \times 10^6$ km³

Greenland Ice Sheet:

Area	$= 1.71 \times 10^6$ km^2
Average thickness	$= 1.725$ km
Quantity of ice	$= 2.95 \times 10^6$ km^3.

Glaciers and Ice Caps:

Area	$= 0.64 \times 10^6$ km^2
Average thickness	$= 0.3125$ km
Quantity of ice	$= 0.2 \times 10^6$ km^3

Total quantity of ice locked in the entire polar surface

$$= (29.0 + 2.95 + 0.2) \times 10^6 \text{ km}^3$$
$$= 32.15 \times 10^6 \text{ km}^3$$

Total surface area of the earth $= 5.1 \times 10^{14}$ m^2

If entire ice in the polar regions melts due to global warming, the approximate sea level rise can be calculated as per the following basic concept:

Total volume of ice all the ice sheets:

$$= 32.15 \times 10^6 \text{ km}^3$$
$$= 32.15 \times 10^{15} \text{ m}^3$$

Total surface area of earth $= 5.1 \times 10^{14}$ m^2.

Height of water level rise if all the water from the ice comes to the total earth surface

$$= \frac{32.15 \times 10^{15} \text{ m}^3}{5.1 \times 10^{14} \text{ m}^2}$$

$$= 63.03 \text{ meter}$$

surface area covered by sea on the earth is 70%

Hence sea level rise = 63.03/0.7 = 90 meters

If all the ice in the polar region melts, it will accumulate in the sea and hence the sea level will rise by another 90 meters. If sea level rises by 90 meters, then most of the world will be submerged under water. In a situation like that only the top portion of few mountains will be exposed above water.

3. Mining of Ground Water

Groundwater depletion will soon be as important factor to the contribution of sea-level rise as the melting of glaciers.

Large quantity of fresh water next to glaciers is stored in the ground. The water stored in the ground can be compared to money deposited in the bank. If we withdraw money from the bank account at a faster rate than our deposits then we will be in trouble. In the same manner, if we use groundwater unsustainably, there might not be enough groundwater available for food production in future. The water pumped out of the ground for irrigation, industrial uses, and even for drinking must go somewhere after it's usage. It may run directly into streams and rivers or evaporates and falls elsewhere as rain, finally ending up in the ocean, thereby raising sea levels further and also resulting in conversion of pure form of water (ground water) to become impure and salty sea water. Hence conservation of ground water is very much important. When the quantity of mining the ground water exceeds the limit of percolation, then ground water level will come down. In that way the quantity of depleting ground water ultimately reaches the sea and cause sea level rise.

Effects of Sea-Level Rise:

Rising of sea levels pose a major threat to island nations and coastal areas. These areas could be swamped and submerged by sea water. Sea level rise will impact all coastal areas, some islands will disappear entirely with even modest increase of sea level by 1 meter. About 23% of the world's people live within 100 km of coastal areas. It is estimated that about 145 million people live within 1 m of mean height above the sea level and 268 and 397 million people live within 5 and 10 m above sea level respectively. Sea level increase will lead to land erosion, habitat destruction in coastal areas and salt water intrusion into the ground water.

On a global scale, the costs of coastal protection and relocation of people from the areas affected by rising sea levels is estimated to be around 200 billion US dollars for an increase in sea level of just 0.5 meters and about 2 trillion US dollars for an increase of 2 meters. In Indian money, the cost will be around 120 trillion rupees (120×10^5 crore). so the money required for the protection (construction of barrier, wall, etc) or relocation after affected by sea level rise is much higher than what we gain by burning fossil fuel. This is to be taken into account at the usage of fossil fuel.

Conclusion:

Burning of fossil fuels, deforestation, urbanisation, destruction of water bodies and over mining of ground water will lead to sea level rise, whereas afforestation, usage of renewable energy sources, cleaning earth's surface area and saving the water bodies (lake, pond, river, stream etc.) will decrease the sea levels and build up the height of polar ice. Our aim should be increase the height of polar ice cap but not the sea levels.

> பழிமலைந்து எய்திய ஆக்கத்தின் சான்றோர்
> கழிநல் குரவே தலை.

> Than store of wealth guilt – laden soul obtain,
> The sorest poverty of perfects soul is richer gain.

நிறைய தவறுகள் செய்து, பின்வரும் நிகழ்வுகளை ஆராயாமல் தன்னுடைய தற்போதைய சுயலாபத்திற்காகச் சம்பாதித்து சேர்த்த சொத்துக்களை விட அறிவுடைய சான்றோரின் வறுமையே மேலானது.

A knowledgeable savant's poverty is better than the wealth of a person who earns it unlawfully without predicting the bad effects but only considering his own benefits

> "Sea level rise is invisible tsunami"
>
> – Stef McDonald

18 UNSTABLENESS OF MATTER

நில்லாதவற்றை நிலை என்று உணரும்
புல்லறி வாண்மை கடை.

To believe that temporary things will last forever
Is deplorable stupidity.

உலகில் அனைத்து பொருள்களும், செல்வங்களும், நாம் காண்கின்ற யாவும் நிலையற்றவை. காலம் மாற மாற அனைத்துப் பொருள்களும், செல்வங்களும், அதன் தன்மைகளும் மாறிக்கொண்டே இருக்கும். இன்று நம்மிடம் இருக்கும் செல்வம் நாளை, வேறு ஒருவரிடம் சென்று சேரும். யாரும் எந்தப் பொருளையும் சொந்தம் கொண்டாட முடியாது. இந்தப் புறப் பொருள்களையெல்லாம் நிலை என்றெண்ணி, எல்லாம் எனக்கே சொந்தமெனும் அறிவின்மை, எல்லா அறிவின்மையையும் விட மோசமானதாகும். இந்த புறப்பொருள்களான செல்வங்களை மட்டும் திருவள்ளுவர் கூறவில்லை. நம் உடலில் இருக்கும் உயிர் வாழத் தேவையான அனைத்து முக்கிய அத்தியாவசியக் காரணிகளாகிய நீர், ஆக்சிசன் மற்றும் கார்பன் போன்ற தனிமங்களுக்கும் இவை பொருந்தும். இவை மூன்றும்- நிலையாக இல்லாமல் மாறிக்கொண்டே இருக்கும், அல்லது இயக்கத்திலேயே இருக்கும். இன்று நம் உடலில் உள்ள ஒரு நீர் மூலக்கூறு நாளை வேறு ஒருவர் உடலில் இருக்கும். அதேபோன்று ஆக்சிசன், கார்பன் மற்றும் அத்தியாவசியமான உலோக மூலக்கூறுகளும், இன்று நம் உடலில் இருந்து நாளை வேறு ஒருவரின் உடலில் சென்றடையும். அதேபோன்று இந்த மூலக்கூறுகள் உயிர் கோளத்தில் இருந்து வாயு கோளத்திற்கும் வாயு கோளத்திலிருந்து நீர் கோளத்திற்கும் மாறிக்கொண்டே இருக்கும்.

This kural denotes that all the materials, wealth and whatever we see in the earth are not permanent. All are temporary. They change its position, size, value and everything from time to time. If we acquire something today, then it will be someone else's tomorrow. No one can claim as it as their permanent property. It is ridiculous if anyone thinks that this kind of changeable asset as their permanent wealth. Thiruvalluvar not only mentioned about the external things but indirectly mentioned internal things including our bodies and their constituents like water, oxygen, carbon and minerals such as nitrates, sulphates, phosphates, etc.

As mentioned about the internal things, our body constituents also change continuously. Our body is made up of oxygen, carbon, hydrogen, nitrogen and other essential minerals. All these water, oxygen, carbon etc are keep on moving continuously from one body to another and from one place to another.

Let us see in detail how it changes its position from one place to another and from one body to another body.

Constituents of Our Body Tissues Are:

Oxygen	65.0%
Carbon	18.0%
Hydrogen	9.5%
Nitrogen	3.2%
Calcium	1.5%
Phosphorus	1.0%
Potassium	0.4%
Sulfur	0.3%
Sodium	0.2%
Chlorine	0.2%
Magnesium	0.2%
Others	<1.0%

Almost 99% of the mass of the human body is made up of six elements such as oxygen, carbon, hydrogen, nitrogen, calcium, and phosphorus. Only about 0.85% is composed of another five elements such as potassium, sulfur, sodium, chlorine, and magnesium. All these 11elements are necessary for life. The remaining elements are trace elements. The mass of all these trace elements put together contributes to less than 10 grams in our body.

These elements will not permanently stay years together in our body. At present the available atoms and molecules in our body tissue will move from our body into the atmosphere by the process of anabolism and metabolism. Again the same atoms and molecules will enter to others body. Hence we cannot even claim our body as our own property.

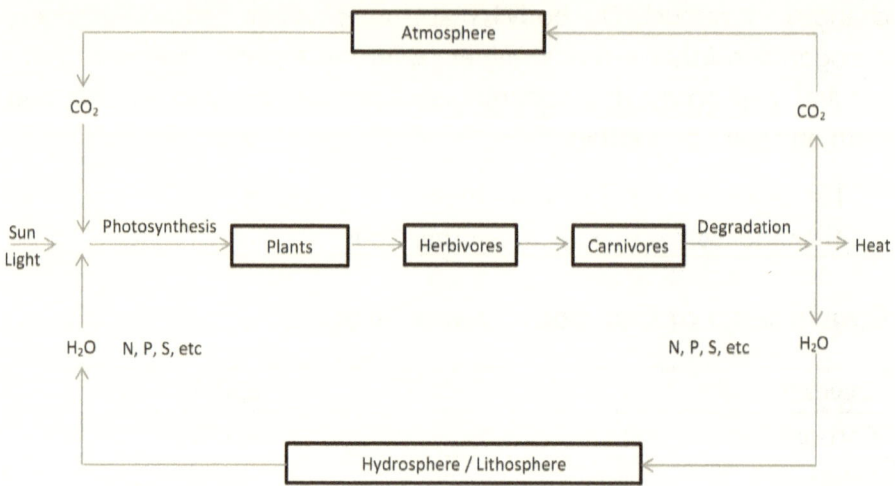

Carbon Cycle:

From the above figure, we can conclude that carbon in the atmosphere as CO_2 enters into the plants through photosynthesis. Then the carbon moves from plants to herbivores like goat, deer, bear, etc then herbivores to carnivores like wolf, fox etc and then further transferred to high level carnivores like lion, tiger, etc. Finally the body of lion and tigers are decomposed by the micro organisms after their death and release carbon dioxide into atmosphere. The released carbon as carbon dioxide again re enter into atmosphere, where it is available for re entry to biosphere.

All atoms and molecules available in our body are received from the atmosphere, Geosphere and hydrosphere. Temporarily they stay in our bodies and change from one sphere to another and also from one body to another body. For example the carbon atom or any other atom which is present in our body today might have been present in the body of a dinosaur, some million years ago. So interestingly, the carbon which were present in the bodies of the king Samrat Ashoka, the Chola king Rajaraja cholan, Gandhiji and Nehuruji might be present in our bodies today.

Oxygen Cycle:

Oxygen enters into the biosphere through water. During photosynthesis, CO_2 and H_2O together synthesise into glucose ($C_6H_{12}O_6$) and oxygen. The evolved oxygen is taken again while breathing by the bio organisms and digest glucose into CO_2 and H_2O. 22% of oxygen by volume present in the biosphere that exists primarily as a component of organic molecules ($C_xH_xN_xO_x$) and water molecules.

(Oxygen concentration in the atmosphere is same as in the biosphere that is the reason our ancient people used to tell as "அண்டத்தில் உள்ளதே பிண்டத்திலும் உள்ளது" means what is there in the atmosphere is the same as what is in the body)

Cycles of Nitrogen, Sulfur, Phosphorus, and Other Nutrients:

Nitrogen, phosphorus, sulfur and other nutrients are very much essential for all bio organisms. These nutrients are being taken through roots by the plants along with water and subsequently they get transferred to all animal bodies through food cycle.

After the death of plant and animals their bodies are degraded by the micro organisms and release the nutrients again into the soil. In such a way, all essential nutrients and minerals continuously change their position from lithosphere to biosphere and subsequently to all spheres.

Conclusion:

We can not claim anything as our own property, not only our wealth, even the constituents of our body also changes continuously from one organism to another. The nutrients and elements that are present in our body today will transfer to another person by natural processes.

So all the constituents in the Earth are not permanent, continuously they get changed from one place to another place and from one person to another person. Hence it is better not to keep more attachment on anything in our life. If we forego attachment on properties and stop being selfish, then we will be honest and happy. All the things available in the earth are common to all.

பொருளானாம் எல்லாம்மென்று ஈயாது இவரும்
மருளானாம் மாணாப் பிறப்பு.

Thinking Wealth is everything and thus, clinging to it and being stingy,
Is a delusion and thus it ll be considered as a wasteful, inglorious life.

இந்த நிலையில்லாத உலகில், நிலையில்லாத பொருட்களின் மீது அளவுகடந்த பற்று வைத்து எல்லாம் எனக்கே சொந்தம் என்று தானும் உண்ணாமல், பிறருக்கும் கொடுத்து உதவாமல் வாழ்கின்றவனுடைய பிறப்பு இழிந்த பிறப்பாகக் கருதப்படும்.

A person who keeps more affection on materials that are not permanent in this world and clinging to them without giving to others is called useless and his birth is considered as most worthless and meaningless.

> "There is nothing so certain that it cannot vary. Even the sun itself has its cycle of unsteadiness. Likewise, there is nothing so mutable that it cannot be fixed. Every revolution produces a new order. Every death is, simultaneously, a metamorphosis."
>
> — Jordan B. Peterson.

19 | ADVERSE EFFECTS OF FOSSIL FUEL BURNING

நல்லார்கண் பட்ட வறுமையின் இன்னாதே
கல்லார்கண் பட்ட திரு.

Not even the poverty of the good is as pernicious as
The wealth that has fallen on the ignorant.

அறிவில்லாதவனிடம் சேர்ந்த, மிகுந்த செல்வமானது கற்றறிந்த நல்லவர்களுடைய வறுமையைக் காட்டிலும் மிகவும் கொடுமையானது. அறம் இல்லாதவனிடம் மிகுந்த செல்வம் சேரும் பொழுது அவன் தன்னுடைய சுயநலத்திற்காகவும், ஆடம்பர வசதிகளுக்காகவும் பல தேவையற்ற செயல்களைச் செய்து தன்னையும் சுற்றுப்புறத்தையும் சீர்கெடுத்துவிடுவான்.

கரிம எரிபொருள் என்ற செல்வம், மனிதகுலத்திற்கு கிடைக்காமல் போயிருந்தால் கூட, மனிதன் தனது சிந்தனையை வேறுவகையில் செலுத்திப் பல புதுப்பிக்கத்தக்க எரிபொருளின் பயன்பாட்டை கண்டறிந்து மனித குலத்திற்குத் தேவையான ஆற்றலை உற்பத்தி செய்து தன்னிறைவு கண்டிருப்பான். புதுப்பிக்க முடியாத கரிம எரிபொருள், எளிதில் அதிக அளவில் கிடைத்ததனால் அவற்றை அளவுக்கதிகமாக எரித்து பூமியை வெப்பமயமாக்கி வாழத்தகாத இடமாக மாற்றிக் கொண்டிருக்கிறான்.

The wealth of an immoral person will give so many ill effects to him as well as others in the society and to the environment than poverty of a wise men. If more wealth is accumulated in the hands of immoral, cruel and selfish person, he will do unwanted and harmful things for his own comfort without considering others.

Here the same can be compared with fossil fuel usage. The fossil fuel is the ancient property to the entire mankind. Now the fossil fuel is used in a unsustainable manner without considering present and future effects to the society. Developed countries like US, Canada and Australia are using more fossil fuels than the developing countries like India and Pakistan. United States is emitting 6.5 metric ton of carbon dioxide per capita per year, whereas India is emitting 1.7 metric ton of carbon dioxide per capita per year. But adverse effects of fossil fuel usage such as global warming, acid rain, water and air pollutions are common to all, irrespective of developed countries or under developed countries.

Prior to the fossil fuel usage, the ancient man during the period of 18th century used renewable source of energy such as watermill and the windmill for pumping water, milling flour and sawing wood.

Fossil fuels are non-renewable energy source but unfortunately it is the dominating source of energy now. More than 82% of the total power usage in the world is fulfilled by fossil fuels. They have variety of application from electricity production to wide range of transport. Moreover fossil fuels are necessary for the production of a variety of common products such as paints, detergents, polymers, cosmetics and some medicines. At present we cannot imagine the modern society without the usage of fossil fuels.

In spite of all development by fossil fuels and its major energy contribution to the entire globe, there are very undesirable negative impacts such as global warming, acid rain, ocean acidification and oxygen depletion. These negative impacts fast track the march towards the extinction of life from the Earth.

Major Problems of Fossil Fuel Usage:

- Global warming
- Ocean acidification
- Oxygen depletion
- Air and water pollution
- Health problem

Global Warming:

Increasing average earth surface temperature is called global warming. Before industrialisation, earth surface average temperature was

+15°C Now the rate of increase of average global temperature is 0.13°C per decade. It is mainly due to the increase of carbon dioxide concentration in the atmosphere by burning fossil fuels. Even though all other factors like solar intensity, angle of radiation and the distance between the sun and the earth are constant, the only major phenomena of globel warming is increasing level of carbon dioxide concentration in the atmosphere. Carbon dioxide concentration in the atmosphere in the pre industrialisation period was 275 PPM, now it is increased beyond 400 PPM. Fossil fuels are not only emitting carbon dioxide they are also emitting sulphur oxides, oxides of Nitrogen and some heavy metals.

Total Global Quantity of Fossil Fuels –

Coal	1139 billion tons
Natural gas	187 trillion cubic metres
Crude oil	1707 billion barrels

1. Main Impacts of Global Warming

1. Sea level rise
2. Releasing the Methane from the polar ice caps
3. Disturbances in the ecosystem

Sea Level Rise:

Global warming leads to numerous ill effects to the environment. Among them one of the most dangerous thing is Sea level rise. Since the beginning of 20th century, between 1900 and 2016 the sea level rose by 16 to 21 cm. Thermal expansion of seawater due to global warming contributed 42% of sea level rise whereas the remaining 58% of sea-level raise is due to the melting of glacier ice sheets. In 2014 the world Meteorological organisation reported that the sea level rise accelerated to 0.12 inches that is 3 millimetre per year on average worldwide. If Greenhouse gas emissions remains unchecked, the Global sea level would rise as much as 3 feet that is 0.9 metre by 2100.

If all the ice that currently exist on the Glacier and ice sheets melts, it would raise sea level by 216 feet. That could cause the entire countries in the world to disappear under the water.

Already 97.5% of water remains in the earth as saline water as it is in the ocean. Only 2.5% of freshwater available in the earth for main purposes like drinking, irrigation, industrial etc. If the sea level rises like this, then all the fresh water from the Glacier and the other water bodies like Lake, rivers and Ponds will be sent to the ocean.

The sea level rise indicates that there is an imbalance between the ice formation in the Glaciers and runoff water from the Glaciers towards the ocean by melting. Always the quantity of water in ice formation must be matched with the quantity of runoff water to the ocean. When run off exceeds, the sea level will rise. When ice formation exceeds,the level of polar ice cap will raise. Sea level rise indicate global warming and the ice cap level rise indicate global cooling. Our all activities should be towards the raising of polar ice cap and not towards raising the sea level.

More than 40% of the total population of the earth are living in the coastal areas. Most of the cities are located in the coastal region. As the rising sea level increases further, the saline water will seep into underground. So the underground water source, soil water, well water and aquifers will become salty due to the intrusion of seawater. Hence lot of people need to be evacuated from the coastal region due to the invasion of seawater into the residential areas. As the rising ocean also erodes the shore line and floods the areas in which coastal animals like coastal birds and the sea turtles are present, those birds and animals will also suffer. The tourism and real estate industry in coastal areas will be affected more.

Releasing of Methane from the Ice Caps (Methaneclathorate):

Methane clathrate (CH_4. 5.75 H_2O or CH_4. 23 H_2O)

Methane clathrate is one among the fossil fuel sources which was developed during ancient period in the polar region. In the early period, the biomass that trapped in ice got anaerobic degradation and produced methane.

$$C_6H_{12}O_6 \xrightarrow{\hspace{4cm}} CH_4$$

(Anaerobic degradation)

Methane produced in this way stayed there for a long time, trapped by a permanent cover of polar ice. Enormous amount of Methane hydrate have been found under the Arctic and Antarctic ice sheets. The total amount of Methane estimated in the region is roughly 10^6 giga ton.

This methaneclathorate is very stable in low temperature and high pressure. As long as the temperature and pressure remains unchanged, it will be stored comfortably in the cage surrounded by ice. If the temperature is increased due to global warming, then the ice cage of the clathrate will melt and plenty of methane will be released into the atmosphere. This released methane will enhance greenhouse effect heavily since the global warming potential of methane is 32, that means 1.0 kg of Methane released into atmosphere is equal to 32 kg of carbon dioxide released to the atmosphere.

Disturbances in the Ecosystem:

The adverse effects of global warming is expected to collapse the entire ecosystem. The plants and animals which are living in the equatorial region will be exposed to higher temperatures due to global warming. Hence they will be forced to move towards polar region in search of conducive temperature and their survival. Organisms already living in that area will be forced to shift further towards polar regions. This will become a very huge problem when the rate of climate change is faster than the rate at which many organism can migrate. Because of this many animals may not be able to withstand in the new climate in that region and may go extinct. If the global warming is not stopped or reduced then one half of the earth plants and one third of the animal species will disappear in the near future.

2. Acid Rain and Ocean Acidification:

Acid rain is any form of rain with acidic components such as oxides of Sulphur and oxides of Nitrogen that falls to the Ground from the atmosphere. This can include rain, snow, fog and hail. Major source of oxides of Sulphur and oxides of Nitrogen are coming from the the usage of fossil fuels.

Adverse Effect:

Acid rain causes so many adverse impact on forests, freshwater bodies and the soil. The lower PH in surface water that occur as a result of acid rain can cause damage to fish and other aquatic animals. At pH lower than 5 most of the fish eggs will not hatch and lower pH can even kill adult fish. As lakes and rivers became more acidic, then the biodiversity in that area will be drastically reduced. Soil biology and chemistry are seriously getting affected by the acid rain. The microbes in the soil will not tolerate at the low PH. Due to this acid rain, agriculture will be seriously affected.

3. Oxygen Depletion:

Fossil fuel burning causes more oxygen depletion. When one mole of carbon burns, it requires one mole oxygen. Oxygen is the most crucial atmospheric component for all living organism in the earth. Oxygen concentration depends on a balance between the processes which emits oxygen such as photosynthesis by green plants and by the processes that consume oxygen such as respiration, combustion and decomposition. Because of the rapid development of industrialisation and modern civilization, especially due to fossil fuel burning, the concentration of atmospheric oxygen has been on a decline over the past 30 years.

Oxygen consumption during the period before 1900 was only 2.0 giga ton per annum. It has been raised 38.2 giga ton per year now. If we do not control the usage of fossil fuels and don't take proper measure for encouraging renewable source of energy, then the oxygen consumption will be 100 giga ton per year in 2100. And the O_2 concentration will decrease from the current level of 20.946%. Human activities have caused Irreversible decline of atmospheric oxygen. It is time to take actions to promote oxygen production and reduce oxygen consumption.

$$C + O_2 \longrightarrow CO_2$$
(12 gram carbon require 32 gram of oxygen.)

Fossil fuel burning per year at current rate is 5.8 giga ton and Oxygen depletion per year is approximately 15.5 giga ton, because 12 gram of carbon requires 32 gram of oxygen to form carbon di oxide

Air Land and Water Pollution:

Fossil fuel usage is not only emitting carbon dioxide, it is also emitting oxides of sulphur, oxides of Nitrogen and heavy metals such as lead and Mercury into the atmosphere. The process of mining fossil fuels also causes so many undesirable chemicals emission into the atmosphere from the interior of the earth. These undesirable chemicals causes so many ill effects to the human health such as asthma, bronchitis, neurological defects and cancer.

The fossil fuels gives many adverse effects than the positive benefits.

இதைத்தான் பாரதியார்

"கண்ணிரண்டை விற்று சித்திரம் வாங்கினால்
கைகொட்டிச் சிரியாரோ"

என்று தமது கவிதையில் கூறியுள்ளார். எந்த ஒரு அறிவுள்ள மனிதனும் தனது இரண்டு கண்களை விற்று அழகான சித்திரத்தை வாங்க மாட்டான். அதுபோலவே மனிதன் தன்னுடைய சுயநல வாழ்விற்காக பூமிக்கடியில் இருக்கும் புதுப்பிக்க இயலாத கரிம எரிபொருளை, அதிக அளவில் பயன்படுத்தி பூமியையே பயன்படுத்த முடியாத அளவிற்கு சீர்கெடுத்துவிட்டான்.

If a person wants to purchase a beautiful picture by selling his two eyes, it wont be considered as wise activity. Only a foolish person can do such type of lunatic activities. Here we can apply the usage of fossil fuels as the beautiful picture. Fossil fuel usage gives a comfortable life now but in future we cannot even exist in earth.

> *"Earth provides enough to satisfy every man's need, but not every man's greed."*
>
> — Mahatma Gandhi

20 CLEANLINESS

தூய்மை என்பது அவாஇன்மை மற்றது
வாய்மை வேண்ட வரும்.

Decrease of desire is purity;
Rest will be obtained by yearning for truth.

தூய்மை என்பது பேராசையற்ற தன்மையாகும். ஆசையை முற்றிலும்
துறந்துவிட்டால் தூய்மை நிலை வந்துவிடும். மனத்தூய்மை,
உடல்தூய்மை, புறத்தூய்மை இவை அனைத்தும் இல்லாமல்
போவதற்கு மூலகாரணம் ஆசையே.

Cleanliness is nothing but eradicating the greed. Greed is the main cause for the degradation of our nature and Environment. If there is no greed, then people will live a simple and happy life with the things they have but greed is not allowing them to do so. It encourages more vehicle usage, building more houses, exploiting our natural resources like land, water, air, soil, etc. All the aforesaid activities cause severe destruction of our Environment. It is all due to the greed of a person or a group or a country.

If we consume resources within the capacity of nature's ability to regain or regenerate, it may not be a problem to the environment. But when we exploit more of natural resources for the fulfillment of our unwanted desires, then the nature will lose it's capacity of regeneration.

Due to depletion of natural resources, getting clean food, clean water and clean air will be the problem for all living organisms. Any nation could not say proudly as a developed country without providing the basic needs such as clean air for breathing, pure water for drinking and healthy food for eating. Hence the preservation of these resources should be the first and foremost priority for a country. In addition to that, the widespread growth of diseases and disruption of natural ecosystems will spoil human health and prosperity. If we do not control the degradation on time, then the losses could not be reversible.

Over exploitation of natural resources will also result in generating huge amounts of waste materials. If we do not take proper care on waste management, then it will pose lot of health problems. Quantity of waste generation will differ based on the life style of the people.

The quantity of waste per capita per day in developed countries like Canada and USA are 2.58 kilogram and 2.33 kilogram respectively. These quantities are huge compared to average Indian per capita of waste generation of only 450 grams. This data proves that the developed countries are using more resources than the actual requirements purely out of their greed.

Quantity of waste generation is a measure of environment degradation. Limiting the waste generation is better for a clean environment as it protects the natural resources. It gives peaceful happy life. But an extravagant lifestyle leads to more depletion of natural resources and hence many troubles and unhappiness.

> யாதனின் யாதனின் நீங்கியான் நோதல்
> அதனின் அதனின் இலன்.

> Whatever thing, a man has renounced, by that thing,
> he cannot suffer pain.

ஒருவன் எந்தெந்த பொருளிலிருந்து பற்று நீங்கியவனாக இருக்கின்றானோ, அந்தந்தப் பொருளால் அவன் துன்பம் அடைவதில்லை.

One who is not greedy of any items and renounced those things then he need not suffer by those properties.

ஆசையே அழிவிற்குக் காரணம்

Desire is the cause of destruction.
        – கௌதமபுத்தர்.

PART 2

பரந்த இவ்வுலகில் சுற்றுச் சூழலியல் பற்றி உலகப் பொதுமறையாம் திருக்குறள் அருளியவற்றை தகுந்த சில உதாரணங்களுடன் விளக்கிட முனைந்தோம். இனி வரவுள்ள பாகத்தில் இடம்பெறும் விளக்கங்களை ஒற்றை குறள் வழி நின்று தெளிவுபடுத்திட எண்ணிய தருவாயில், திண்ணமாய் சான்றோர் பலரால் தெளிவுரை வழங்கலான குறளாக நிற்பது அறிவுடைமை அதிகாரம் குறள் எண்: 423 ஆகும்.

எப்பொருள் யார்யார்வாய்க் கேட்பினும் அப்பொருள்
மெய்ப்பொருள் காண்ப தறிவு

இக்குறளின் விளக்கத்தை பல்வேறு விதமாய்க் கூறக் கேட்டாலும், அறிவார்ந்த செயல் எனப் போற்றப்படுவது கூற்றை நோக்கித் தெளிவு பெறுவது அல்லது கற்றோன் செயலாக இருப்பது, சொல்லப்படும் கருத்தின் மெய்த்தன்மையைக் காண்பதாகும். கூறப்படும் கருத்தின் மெய்த்தன்மையைக் காண்பது பலரிடமிருந்து பல்வேறு கருத்துக்களை பெற்று பின் முடிவாய் உன் அறிவைக் கொண்டு ஆராய்ந்து முடிவெடுக்க வேண்டுமென்பதேயாகும்.

பலதரப்பட்ட கருத்துக்களை பலரிடமிருந்து பெற்று அதன் மெய்த்தன்மையை காண்பது அறிவு எனவும் ஒரே கருத்தை பலவாறு பலரிடமிருந்து பெற்று பின் முடிவெடுப்பது அறிவுடைமை எனவும் வள்ளுவம் சொல்கிறது. ஒரு பானை சோற்றில் ஒரு சோறு பதம் என்பது பழமொழி. ஆக இனி தரவுள்ள பகுதியில் கூற உள்ள அனைத்து விளக்கங்களயும் உள்வாங்கி இக்குறளின் கூற்றுக்கேற்ப மெய்ப்பொருள் காண்போமாக.

In this diverse world, how the concepts of environmental science had been touched in Thirukural was seen clearly with some of Thirukural highlights in the previous chapters. Hereafter we are going to see some discussions which will be highlighted with a single Thirukural for which a lot of explanations were given by various prominent scholars. A Kural in Athikaram Arivudamai (The possession of knowledge)

எப்பொருள் யார்யார்வாய்க் கேட்பினும் அப்பொருள்
மெய்ப்பொருள் காண்ப தறிவு

Wisdom is in deep inquiry of words and intentions
Not in trusting the lips that may utter for deception

Even though there were various explanations for this Kural, such as "a knowledgeable person clarifies the truth without seeing the person who delivers" or "the literate people look at only the truth of any concept." One more explanation would be to "get a lot of suggestions and apply your sense and pick the truth among them." These are the explanations offered by the philosophy of valluvam. A tamil Proverb asserts that "a single grain of rice can be taken as a sample for the entire pot of rice to ascertain the state of its cooking" Like wise with this single Kural in mind, we can go through the forth coming chapters

1 WE ALL KNOW THAT ONE DAY ON EARTH IS 24 HOURS. DO YOU THINK IT IS ALWAYS CONSTANT?

No, The Earth's rotation is being slowed down by the presence of the Moon. As Earth rotates, the Moon's gravity causes the pulling of ocean water towards it, which results in rise and fall of sea levels (tides). This interaction of moon over the Earth causes little bit of friction between the tides and the rotating Earth. This friction effect causes the rotation of the earth to slow down slightly. As Earth slows, the Moon is being pushed away slightly from the earth.

Every year, the Moon gains a little energy from the Earth, and drifts a little farther away. The moon was previously nearer to Earth in the earlier times than today. It was much closer to Earth during its formation.

Both the drifting away of the Moon and the friction between the ties and earth's rotation, causes deceleration of the earth's speed. However this effect is very meagre. The slowing of the Earth's rotation over the last 100 years is estimated to be about 1.4 milliseconds.

That's a deceleration of 0.0014 seconds in total, over 100 Years.

If we calculate the period of time required for 1 second deceleration of the speed of the earth, it ll be 71428 years.

In 100 years, the earth's speed for one rotation decreases by 0.0014 second, hence the time required for 1 second slows down

$$= \frac{100}{0.0014} \times 1$$

$$= 71428 \text{ years}$$

For increasing earth's day by 1 hr i.e from 24 hrs. to 25 hrs., the time required will be

$$= \frac{71428}{1} \times 3600$$

= 257140800 years (257 million years)

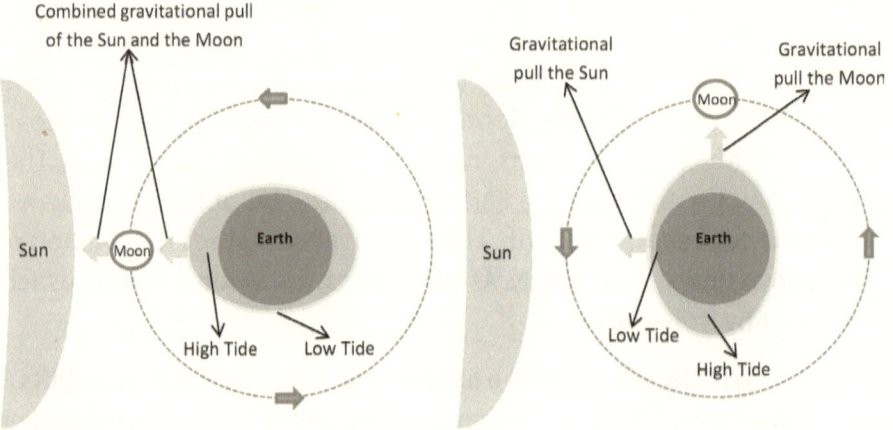

As per this calculation, after 257 million years, one day of the earth will be 25 hours. Billions of years ago, when moon was formed, one Earth day was only 5 to 6 hours.

This activities of Earth and moon's interaction on the rotation speed of earth will come to an end when the length of the earth day is the same as the duration of moon's rotation. Interestingly this will happen when the Moon takes 45 days to orbit the Earth and the Earth takes 45 days to complete one rotation. At that point the earth will always show the same face to the moon, as the moon already does to us.

> *"Time and tide wait for no man"*
>
> — *Geoffrey Chancer*

2 CAN OXYGEN BE TOXIC?

Yes! higher oxygen level in the atmosphere is harmful to humans as well as to the environment. We all know that oxygen is a very important component for all the living beings on the earth and it is essential for survival. If the same oxygen increases beyond a particular limit, then it becomes harmful and toxic.

This example is a proof of the Tamil proverb "அளவுக்கு மிஞ்சினால் அமிர்தமும் நஞ்சு" which means Too much of anything is good for nothing even if it is a precious thing.

Safe Oxygen Levels

The Occupational Safety and Health Administration (OSHA) determined that the optimal range of oxygen in the air for human health is between 19.5 and 23.5%. The normal atmospheric air contains approximately 78% of nitrogen and 20.9% of oxygen. The remaining fraction is made up of primarily argon and trace amounts of CO_2, neon, water vapour and helium.

The oxygen concentration in the atmosphere should be between 19.5% and 23.5%. When oxygen concentration drops below 19.5% our body cells fail to receive the oxygen needed to function correctly, and it will lead to hypoxemia. This hypoxemia will cause many complications to our body tissues and organs.

On the other hand, higher concentration of oxygen than normal level is also not suitable for human health as well as for the environment.

Adverse Effect of Higher Concentration of Oxygen in the Atmosphere:

When the concentration of oxygen exceeds beyond 23.5% it will causes dangerous effects to human health. Slightly higher-than-normal oxygen levels aren't so harmful to life. But extremely high concentrations of oxygen in the air causes harmful side effects to humans. Very high level of oxygen causes oxidization of free radicals in our body. These free radicals attack the tissues and cells of the body and cause muscle twitching. The effects of short exposure to the higher oxygen levels can mostly be reversed, but lengthy exposure can lead to death. The higher oxygen levels lead to hyperopia (excess of oxygen in body tissues).

Health effects will vary depending on the type of exposure. Symptoms may include disorientation, breathing problems, and vision changes such as myopia. Prolonged exposure to above-normal oxygen levels or shorter exposures to very high oxygen levels can cause oxidative damage to cell membranes, collapse of the alveoli in the lungs and retinal detachment.

As we breathe in, our lungs send oxygen to the tissues and as we breathe out, they expel CO_2 out of our body. If a person inhales too much oxygen, he will be unable to exhale the CO_2 completely because of more oxygen presence in it even after sending to tissues. This can lead to a condition called hypercapnia. (condition of abnormally elevated CO_2 concentration in the blood)

Adverse Effect of Higher Concentration of Oxygen in the Environment:

Higher levels of O_2 more than normal (20.9%) in the atmosphere will increase the flammability of all combustible matter. Auto-ignition temperature of materials decreases due to the higher oxygen concentration. Materials that do not get ignited in normal conditions may burn in an O_2 enriched environment. The materials which burn normally in air will burn hotter and faster in those conditions. Even a small increase in the O_2 level (to 24%) can create a dangerous situation. In that conditions where there is extreme concentrations of oxygen, fires cannot be put off by any means.

However, there is no chance of increasing oxygen concentration in the atmosphere, either naturally or by human anthropogenic activities

"God is like oxygen. You can't see him,
but you can't live without him."

3 HEAT DEATH OF THE UNIVERSE — THE ULTIMATE DESTINATION

Heat death does not specify any particular absolute temperature as a cut off point below which heat can not be extracted. This phenomena will occur when the universe reaches thermodynamic equilibrium. If there is no heat source and no heat sink then we can not extract energy from heat at any temperature by any means. The 'heat-death' of the universe will happen when the universe reaches a state of maximum entropy. This happens when all available energy (source) has moved to places of less energy (sink). At that time, heat ceases to flow and no more work can be acquired from the heat.

The sun and other stars in the universe are radiating heat into the universe, When these stars completely loose their fuel, the entire universe will attain thermal equilibrium. All celestial bodies in the universe will be at the same temperature.

Entropy:

Entropy is the measure of energy which is not available for work in any system. The higher the entropy, the energy available to do work is less. When the system reaches equilibrium, the entropy reaches a maximum value and the system can not function. The entropy of the universe increases because universe is an isolated system and heat never flows from low temperature to higher temperature. Heat always flows spontaneously from higher temperature to lower temperature causing an increase of entropy. Irrespective of the process, whether it is reversible or irreversible, the net change of entropy is always positive.

If the process is reversible, then the change in entropy is equal to the heat transferred divided by the temperature at which process takes place. Let as consider Q as the heat transferred, T as the temperature, and S as the entropy.

Let us consider mathematically, an amount of energy "q" is transferred from a hot region at temperature T1 to a cold region at temperature T2.

The entropy change S1 of the hot region is defined as

$$S1 = \frac{q}{T1}$$

The entropy change S2 of the cold region is

$$S2 = \frac{Q}{T2}$$

Therefore, during the energy transfer, the net change in entropy is

$$\Delta s = S2 - S1 = \frac{q}{T2}(-)\frac{q}{T1}$$

$$= q \times \frac{1}{T2}(-)\frac{1}{T1}$$

Since T2 < T1 and

$$= \frac{1}{T2} > \frac{1}{T1}$$

ΔS is positive since q is positive.

Far in the future, when all the possible reactions have taken place in all celestial bodies and all the heat were spent, then maximum entropy will be achieved. At the maximum entropy no reaction will be possible, because the universe will have reached thermal equilibrium. The reactions that can cause decrease of entropy is not possible, so in effect the universe will have to die.

This is called as heat death of the universe, also known as the Big Chill or Big Freeze. It is the ultimate fate of the universe in the far, far. ... future.

"If heat death of the universe is the destination, it really is all about the journey" ki.

4 | WHAT IS A GREENHOUSE? ARE THE GREENHOUSE GASES GREEN IN COLOUR?

A greenhouse, also called a glasshouse (Hot house) is an enclosure with walls and roof made up of transparent material, like glass or fibre. These enclosures range in size from small sheds to industrial-sized buildings. The interior of a greenhouse exposed to sunlight becomes significantly warmer than the exterior space.

In some of the cold countries where the temperature is very low, plants cannot grow and sustain their life. For growing plants in such places, warm conditions are required. So glass houses were built in such a way that the interior temperature of the glass houses are higher than the atmospheric temperature.

The warmness in a greenhouse occurs because the incident solar radiation passes through the transparent roof and walls and is absorbed by the floor inside the house, and becomes warm. As the structure is not open to the atmosphere, the warm air cannot escape via convection, so the temperature inside the greenhouse rises.

During ancient times, greenhouses were built in Rome for growing fruits and vegetables. In the first century, portable greenhouse was built with a cover made up of transparent stone. This unusual and rare greenhouse was designed to cultivate the emperor's favourite vegetable, the cucumber. In later stages, cucumbers were planted in wheeled carts which were kept in the sunlight during day time and then taken inside to keep them warm at night.

Now many commercial glass greenhouses are equipped with heating, cooling, lighting provisions for growing vegetables, flowers and fruits. Climate can also be controlled by a computer to optimize the conditions for plant growth in order to reduce risk of production failures. Many vegetables and flowers can be planted and grown inside the greenhouses in late winter and early spring. In summer season they can be transplanted outside as the weather warms. Pests, diseases, extreme cold and extreme humidity can be avoided in the greenhouse plantations. The Netherlands and Canada have some of the largest greenhouses in the world. In Netherland greenhouses occupy 10,526 hectares of land.

Ventilation is also provided at the top of the greenhouse. Ventilation is one of the most important component for a successful greenhouse. If there is no proper ventilation, then greenhouses and their growing plants can become prone to problems. The main purpose of ventilation is to regulate the temperature and humidity to the optimal level, and to ensure movement of air. Ventilation also ensures supply of fresh air for photosynthesis and plant respiration. The enhanced CO_2 level inside the greenhouse will increase productivity.

The incoming solar radiation, having maximum visible and small portion of UV rays, easily penetrate through glass in the roof of the greenhouse due to high frequency and low wave length. The rays which pass through the roof of the greenhouse heats the interior parts. The heated interiors in turn, will emit infrared radiation, which cannot pass through the glass roof to outside atmosphere due to their low energy (low frequency) and also long wave length. Hence heat accumulates inside the greenhouse. Due to this phenomena, inside of the greenhouse is slightly warmer than the outside atmosphere.

Similar to the function of the glass in the greenhouse, few gases in the atmosphere traps the heat, not allowing them to release back to the space, resulting in increasing the temperature of earth. So the gases which are responsible for warming of the globe are named as greenhouse gases and they are not green in colour.

> "The violence that exists in the human heart is also manifest in the symptoms of illness that we see in the Earth, the water, the air, and in living things"
>
> — Pope Francis.

5   IS WOOD A RENEWABLE ENERGY SOURCE OR NON RENEWABLE ENERGY SOURCE?

Wood is a renewable energy source, because it can be produced again within a period of time.

Renewable Energy

Renewable energy is energy produced from sources that do not get depleted, or can be replenished within a reasonable time period. The most common examples for renewable energy are wind, solar, geothermal, biomass and hydropower. Renewable energy resources are inexhaustible.

Non-Renewable Energy:

A non-renewable energy resource is one that could not be replenished within a reasonable period of time. Examples of non-renewable energy sources are coal, oil and natural gas. The time required to produce these sources takes million of years but it is being spent-much more quickly than nature could renew. It will not be available once it is exhausted. Non-renewable resources are also called exhaustible resources.

Wood is growable. Wood comes from trees and trees are easy to grow. Only we need fertile ground, water, and sunlight for growing trees. Therefore it is a renewable resource. For example if we plant trees in a specific area, like in an acre, it ll yield a certain amount, say one ton of wood in a specific period of time, like in about 10 yrs. After harvesting, this process can be repeated over and over indefinitely. Wood cannot be compared with coal, oil and natural gas because they might have required billions of years to form.

Burning of wood releases CO_2, but it has still been classified as a renewable energy source, because during photosynthesis, the CO_2 released during the burning of wood will be captured again from the atmosphere by the new plants and trees.

Environmental Impact:

On combustion, the carbon from biomass is released into the atmosphere as carbon dioxide (CO_2). After a period of time ranging from a few months to decades, the CO_2 released from combustion is absorbed back by new plants or trees.

One interesting thing here is, in this cyclic process of plant (trees) grown for harvesting energy either for power production or some other uses, in each cycle, some amount of biomass (carbon) such as plant roots and waste residue of plants will be left behind after harvest, which will be added to the soil. In that way some fraction of CO_2 in the atmosphere is being transferred into the ground as soil carbon. This carbon in the soil will stay for longer duration as sink. The quantity of soil carbon around the globe which is available within few feet below the earth surface is more than that of atmosphere and all biomass carbon quantity. Hence when soil carbon increases, the atmospheric CO_2 will be reduced. In that manner, the atmosphere CO_2 concentration is being reduced due to this continuous planting of trees.

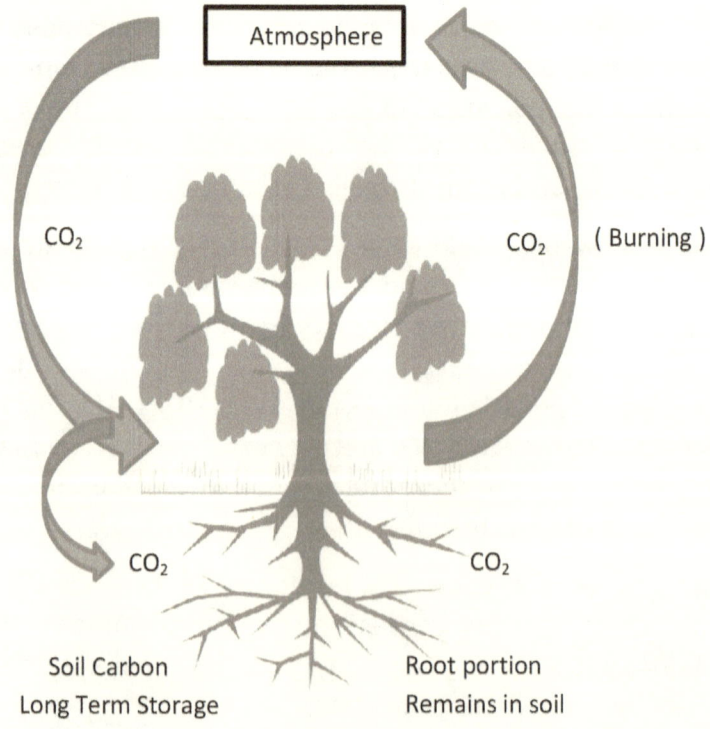

Scientists estimate that carbon residue in soil is more than the quantity in the atmosphere and biosphere combined. There are 2,500 billion tons of carbon in soil, compared with 800 billion tons of carbon in the atmosphere and 560 billion tons in biosphere. (plant and animal biomass)

Dendrothemal Energy:

Dendro power is the generation of electricity by using biomass as a fuel. It is particularly well suited to tropical countries such as Sri Lanka and Philippines as the firewood can be grown rapidly. After cutting the trees for power production, the bottom portion of stem is left in the ground for regrowth. The tree will again regrow shortly on its stem or its root. In this way the root portion of the plants will be accumulated continuously in the soil which increases soil carbon. By this, the atmospheric CO_2 is taken out and stored in the ground as soil carbon.

Energy crops, which are mostly low-cost and low-maintenance crops, are grown solely for energy production (not for food). These crops are burnt to generate power or heat. For producing dendrothemal energy, we have to identify suitable vacant land and woody trees. Energy crops are to be cultivated in that vacant land in cyclic manner for sustainable power production. By encouraging dendrothemal energy, we can reduce coal fired thermal power plants. Dendrothemal energy will decrease the CO_2 concentration in the atmosphere whereas coal fired thermal power plants will increase it.

Conclusion:

We can conclude that wood is also one among the renewable energy sources. But while burning the wood or other biomass for harvesting energy, the surrounding areas will be polluted considerably. If we can develop suitable technology for removal of the soot, ash and CO_2 at the source itself, then the development of dendrothemal energy will be very much effective which will lead to a clean environment.

> "Look deep into nature, and then you will understand everything better"
>
> — Albert Einstein

6   IS CARBON THE WORKING SUBSTANCE OF THE BIO ORGANISMS?

Yes! Carbon acts as working substance between the bio organisms (Plants as well as animals) and the atmosphere. For example refrigerants like ammonia and Freon are engaged as working fluids in the refrigerators and air conditioners. These refrigerants works in cyclic process which will remove heat from the room in order to give cooling effect and eject the heat to the atmosphere. In the same way, water in thermal power plants are used as working fluids. The water receives heat from the boiler and converts it into steam. The steam is used for running the turbine and produce electricity. The condensed and cooled water from the turbine comes to boiler again and it is recycled. In the same way, carbon also act as a working substance between atmosphere and biosphere.

The carbon as carbon dioxide enter into the plants from atmosphere and gets converted into glucose by the process of photosynthesis in the presence of sunlight and water. The carbon atom in the glucose on respiration or by degradation will be split into CO_2 and water. That CO_2 will re-enter to the atmosphere.

During the process of photosynthesis, carbon obtains energy from the sun light and ejects the energy into the bio organisms in its cyclic process.

$$6CO_2 + 6H_2O \xrightarrow{\text{(Sunlight)}} C_6H_{12}O_6 + 6O_2 \text{ (photosynthesis)}$$

(Carbon enters from atmosphere to biosphere)

$$C_6H_{12}O_6 + 6O_2 \longrightarrow 6CO_2 + 6H_2O \text{ (respiration) +Heat}$$
(Carbon coming out from the biosphere to atmosphere)

Carbon acts as working substance

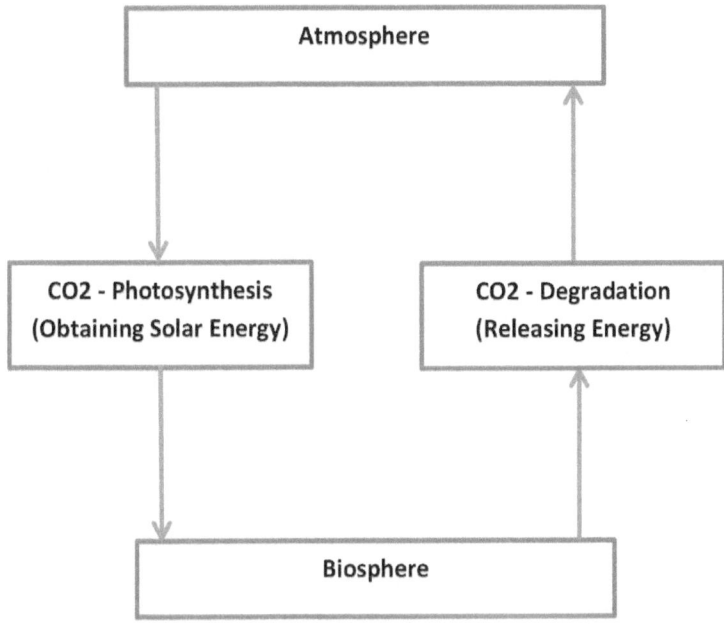

Lives of all bio organisms depends on these two reactions.

The first reaction shows gaining energy from the sunlight and the second reaction shows the energy released as heat into the atmosphere. In these reactions everything (oxygen, carbondioxide, water) are in cycles except solar radiation.

As per the stoichiometry calculation there will be no mass defect for obtaining energy like nuclear reaction. In all these reactions carbon atom in CO_2 gains two electrons during photosynthesis and the same two electrons are released on expiration through degradation or burning.

When Comparing glucose and carbon dioxide (CO_2), carbon dioxide is a low-energy compound. Since CO_2 is in a highly "oxidised" state and glucose, is in a more "reduced." State, the carbon atom in the CO_2 is reduced by obtaining energy from the sunlight and form glucose.

(The oxidation state of carbon can vary widely, from −4 to +4. Highly reduced carbon is in methane molecule (−4) and highly oxidised carbon is in CO_2 molecule (+4). Oxidised state of carbon in the glucose molecule is zero)

Actually CO_2 is reduced during photosynthesis and gains two electrons from water molecules. In that way the inorganic carbon in the CO_2 molecule becomes organic carbon in glucose. Normally the carbon in oxidised state is called inorganic carbon whereas reduced carbon is called organic carbon. Carbon gains two electrons from the process of photosynthesis and it is donated to the bio organism in the form of glucose. Upon degradation of the glucose, the two electrons are removed and converted to CO_2 further.

Interestingly, the entire bio organisms in the Earth from algae to big trees and bacteria to whales including humans exist only by the dancing of these two electrons between inorganic carbon and the organic carbon. Normally people do not like black colour, but the function of this black coloured carbon in the environment and in the biosphere is a significant one.

Only Carbon and boron has the colour black among all the other elements.

"You will die but the carbon will not; its career does not end with you.

It will return to the soil, and there a plant may take it up again in time,

sending it once more on a cycle of plant and animal life."

— Jacob Bronowski.

7 ARE BLACK AND WHITE HOLES MERE HOLES?!****

Black hole is not merely a hole. When the stars in the universe whose mass is higher than the sun collapse due to depletion of fuel and shrink into unimaginably minimum sizes, they became compressed stars, and are called as black holes. If all the fuel in the Sun is depleted (Hydrogen) completely, then the size of the sun would shrink substantially and would form a black hole with a radius of just 3 kilometres.

Actual radius of the sun is $\quad = \quad 695700$ km.

The actual volume of the sun $\quad = \quad 1.41 \times 10^{27} m^3$

Mass of the sun $\quad\quad\quad\quad = \quad 1.9885 \times 10^{30}$ kgs

Average density of the sun 1.408 gm/cc

When the sun gets diminished to a sphere of 3 km radius, then the density of sun is calculated as follows.

Volume of sphere $\quad\quad = \quad \dfrac{4\pi r^3}{3}$

Squeezed volume of sun $\quad = \quad \dfrac{4\pi 3^3}{3}$

$\quad\quad\quad\quad\quad\quad\quad = \quad 113.4$ km^3

The new volume of sun after it is squeezed will be

$\quad\quad\quad\quad\quad\quad\quad = \quad 113.4 \times 10^{15}$ cm^3

The mass of the sun $\quad\quad = \quad 1.9885 \times 10^{30}$ kgs

$\quad\quad\quad\quad\quad\quad\quad = \quad 1.9885 \times 10^{33}$ grams

The density of the sun after diminishing and converted into a black hole will be

1.9885 × 10^{33} grams

$$= \frac{1.9885 \times 10^{33} \text{ grams}}{113.4 \times 10^{15} \text{ cm}^3}$$

$$= 1.795 \times 10^{16} \text{ gram/cc}$$

$$= 1.795 \times 10^{10} \text{ ton/cc.}$$

This 1.795×1^{10} ton of mass will be compacted within a volume of 1 cm^3 during formation as a black hole.

Due to this huge density, the gravitational force will become tremendous and it can suck everything towards it including light.

Similarly, when the stars shrink into very small volumes then their densities will become extremely high and form black holes. In such conditions, with so much mass in a confined volume, the collective force of gravity will pull all the objects towards them. Even light cannot escape from a black hole. Black holes are unbelievably dense matter with a gravitational force millions of times greater than what we experience on Earth.

The closest black hole to the earth is V616 Monocerotis, also known as V616 Mon. It is located about 3,000 light years away from the earth and has mass of 9–13 times higher than the Sun. Billions of black holes are available in the universe.

White Hole:

White holes are opposite of black holes. It could not keep anything within it. Even light will not enter into the white hole. Both Matter and light will escape from it. Like black holes, white holes also have properties like mass, charge, and angular momentum. They attract matter like any other mass, but objects attracted towards a white hole would never actually reach them.

Every action has an equal and opposite reaction. A black hole sucks matter from the universe, but a white hole reject all the things. A white hole looks similar to a black hole, but instead of absorbing matter inside, it pushes the matter out.

> "Black holes are where
> God divided by zero"
>
> — Albert Einstein

8 IS WATER VAPOUR A MAJOR CONTRIBUTOR TO GREENHOUSE EFFECT THAN CARBONDIOXIDE?!

Yes! global warming is happening not only because of increasing CO_2 and methane concentration in the atmosphere. Increasing water vapour concentration in the atmosphere also is a major cause for global warming.

Water Vapor is the most abundant greenhouse gas in atmosphere than CO_2. Water vapour represents about 80% of total greenhouse gases by mass in the atmosphere and 90% of greenhouse gases by volume. Water vapor and clouds account for 66 to 85% of the greenhouse effect depending upon the area. Whereas CO_2 contribution towards greenhouse effect is only 9 to 26%. Hence water vapour plays a significant role in greenhouse effect than any other greenhouse gases including CO_2.

Greenhouse gases like carbon dioxide, nitrous oxides, methane, and ozone are trace gases that totally account to less than 0.01% in atmosphere. whereas Water vapor is unique and its concentration varies from 0–4% in atmosphere.

The following is the order of contributions of various gases to greenhouse effect. Most contribution is by water vapour and least contribution is by HCFC.

Water vapor (H_2O) > Carbon dioxide (CO_2) > Methane (CH_4) > Nitrous oxide (N_2O) > Ozone > Chlorofluorocarbons (CFCs) > Hydro fluorocarbon.

Even though water vapour is a major contributor of greenhouse effect, we have never bothered about water vapour because the atmospheric concentration of water vapor is not in our hand. Any human activities will not increase or decrease the water vapour concentration directly and also it will not stay more than nine days in atmosphere. Once water evaporates, it will return to the earth as rain within nine days.

Depending upon the global temperature, the concentration of the water vapour in the atmosphere will vary. The amount of water vapour that can be hold by one unit mass of air depends on the temperature and pressure of the atmosphere. The dynamic equilibrium of water vapour concentration in the atmosphere will increase as temperature increases. The water vapour holding capacity of air at 30°C is 30.4 gram, whereas at 40°C the holding capacity will be 51.1 gram per kg of air. The maximum sustainable water vapor concentration increases by about 7% for every degree Celsius of increase in temperature.

Clouds too depend on temperature, pressure, convection and amount of water vapor present.

So when there is a rise of temperature in the atmosphere because of external factors, it will lead to increase in water vapour concentration. Since water vapor is a potent greenhouse gas, it causes the temperature to raise further.

Thus, the total greenhouse effect after increasing CO_2 concentration will not only be the effect caused by CO_2 alone, but also the combined effect of water vapour too.

For instance, when CO_2 concentrations are doubled, then the absorption of IR rays by the CO_2 alone would increase by 4 W/m², but once the water vapor and clouds are together the absorption increases by almost 20 W/m². Hence the 16 watt per m² excess increase is caused by the effect of water vapour. Studies show that if there is a 1°C temperature increase caused by CO_2 effect, then the water vapor will cause the temperature to go up another 2°C. The total warming caused by CO_2 together with water vapour will be 3°C.

In a city named Salah in southern Algeria, day time temperature is recorded as 52°C, but at midnight in the same place temperature drops to (–) 3.6°C. Such a low temperature at night is caused due to less greenhouse effect because there is no water vapour in the atmosphere.

It's true that water vapour is the biggest overall contributor to the greenhouse effect and that humans are not directly responsible for it. However rising of water vapour concentration is indirectly caused by man-made emissions of CO_2 and other greenhouse gases in the atmosphere.

Evaporation continuously acts to increase the Earth's greenhouse effects,

... While precipitation continuously acts to decrease greenhouse effects.

9 DOES EARTH WORK AS A HEAT ENGINE?!

Yes! Earth has been working continuously as a heat engine just like any other engine created by us

Difference of temperature between a source and a sink and a working substance are the important elements for a heat engine to produce work.

This can be done through a heat engine in two ways. One is by the change in volume of the working substance by hot and cold reservoirs that makes it to alternately expand and compress. The other one is convection – the change of density of the working substance resulting from the change of volume.

Earth also works like a giant heat engine working between two reservoirs with different temperatures. The hot reservoir is solar radiation and cold reservoir is the outer space. Working substances are the atmosphere (air) and water.

The Earth uses solar power to move air in the atmosphere and water in the hydrosphere. This engine continuously drives the water cycle and air cycle.

Water Cycle:

Water moves from the oceans to the atmosphere by evaporation, from the atmosphere to the land by precipitation, and from the land back to the oceans by the gravitational force. The cycle begins when the ocean and other water bodies absorb sunlight and causes evaporation. This evaporated water vapour travels to high altitudes in the atmosphere and loses it latent heat, condenses and comes as rain. By this hydrological cycle earth is not only getting rain, but also earth's surface heat is taken away in the form of latent heat and transmitted to the sky.

Total quantity of rainfall per year is approximately 1000 mm (1.0 meter) around the globe

Total surface of the earth $= 5.1 \times 10^{14} \, M^2$

(Around the globe 1meter rain fall, hence the volume will be total earth surface multiplied by 1meter)

Total volume of rain water per year $= 5.1 \times 10^{14} \times 1 M^3$

$$= 5.1 \times 10^{17} \text{ lts or kgs}$$

Energy required to evaporate one kg of water from 15°C is 2465 kilo joule.

Energy taken from the solar radiation in one year for evaporation is

$$= 5.1 \times 10^{17} \times 2465 \text{ kilo joule}$$

$$= 12571.5 \times 10^{17} \text{ kilo joule}$$

Energy required for one day
$$= \frac{12571.5 \times 10^{17} \text{ kilo joule}}{365}$$

Energy required for one hour
$$= \frac{12571.5 \times 10^{17} \text{ kilo joule}}{365 \times 24}$$

Energy required for one second
$$= \frac{12571.5 \times 10^{17} \text{ kilo joule}}{365 \times 24 \times 3600}$$

$$= 3986396' \times 10^7 \text{ kilo joule}$$

Energy required per square meter per second

$$= \frac{3986396' \times 10^7 \text{ kilo joule}}{5.1 \times 10^{14} \, M^2}$$

$$= 78 \text{ watt. or 78 joule per second}$$

Solar energy utilised per second per square metre for evaporation (water cycle) is 78 watt.

Total solar energy input per square metre per second is 342 joule

Hence fraction of solar energy utilised for latent heat of evaporation

$$= \frac{78}{342} \times 100$$

= 22.8 percent

Approximately 23%

In the total input of solar radiation, 23% of the energy is utilised by the earth for latent heat of evaporation (water cycle)

Air Cycle:

The earth as a heat engine works by using two different working substances.

One is water and another is air. Atmospheric air act as a working fluid that can generate kinetic energy by transporting the energy from a warm source to a cold sink. In effect, warm air rises, cold air sinks, and wind is generated.

In the atmospheric air cycle, air takes solar heat from the surface of the earth and transmits the heat into the upper atmosphere. By this way 7% of the solar radiation falling on the earth is sent out to the sky. Approximately ($342 \times .07 = 23.9$ joule) 24 joules of solar energy per second per square meter is utilised by the earth for atmospheric air circulation.

The earth as heat engine, works continuously to even out the imbalance created by solar radiation around the entire surface of the earth and atmosphere by creating evaporation, rain and wind.

When the flow of incoming solar energy to the earth is balanced by an equal flow of heat to space from the Earth, then the earth is said to be in radiative equilibrium, and global temperature will be relatively stable. Anything that increases or decreases the amount of incoming or outgoing energy will disturb Earth's radiative equilibrium which will lead to rise or fall in global temperatures.

The earth as a heat engine not only redistribute solar heat from the equator toward the poles, but also from the Earth's surface to the space. Otherwise, Earth would endlessly get heated up and burnt into ash some million years ago. Earth's temperature doesn't infinitely rise because the earth surface and the atmosphere are simultaneously radiating heat to space. This net flow of energy into and out of the Earth system is Earth's energy budget.

Unlike earth, the adjacent planet Mars has not evolved life. Mars is considered to be a poorer heat engine than the Earth. Mars is very smaller than the Earth. Due to less mass and very weak gravitational force prevailing in Mars, atmosphere is also very thin. A planet that does not sufficiently function as a heat engine is considered as a dead planet. No life is possible in that planet.

> "The earth does not belong to us,
> We belong to the earth."

10 IS THERE ANY ANIMAL WHICH LIVES UNDER EXTREME PRESSURE – OVER 600 BARS?!

Yes! Some marine organisms lead their life at 6000 meters depth in the sea floor. These organisms which live in the sea floor are called benthos. The area near the sea floor is called abyssopelagic zone. At 6000 meter depth, the pressure will be around 600 atmospheres (i.e 600 times more than the sea level atmosphere). The organisms live in that zone cannot live at the surface levels of sea (top levels). If these benthos organisms were taken from the depths up to higher levels, then they will explode due to low pressure.

Characteristics of Deep Sea:

In the deep sea, there will be no light hence there is no day and night. Very low temperature and high pressure exist in that area. Due to the absence of light there will be no vegetation. So the animals living there will depend on other animals for their food. The main source of food for those animals are carcasses (decomposed remains of dead organisms) or excreta of the organisms living at top levels of sea which ll sink to the bottom. The benthos will consume them and clean the sea floor. Otherwise the sea floor will be dumped with huge mass of decomposed debris which spoil the entire ocean environment. In that way the benthos are called as scavengers of the sea floor.

Bioluminescence:

Emitting light is a remarkable characteristic of the benthos. They produce their own light from their bodies, with the help of which it will search for food and shelter. This bioluminescent (emitted from a living organism) light is either green or blue in colour. It is commonly called as "cold light" because it will not produce any heat.

The benthos, scavengers of the sea will take care of the clean environment in the sea floor. All the carcasses of marine organisms from micro level to whales will automatically fall into the sea floor. If the benthos are not available in that area and not scavenging the dead bodies and excreta, sea will lose its capacity to absorb excess CO_2 from the atmosphere, the entire ocean ecosystem will be affected, ultimately our atmosphere will also be affected.

In such a way the benthos are helpful to the entire ecosystem and environment both in the ocean as well as land by their natural activities.

> "The world's finest wilderness lies beneath the waves"

11 IS GREENHOUSE EFFECT BAD?

No, Because of greenhouse effect only life exists in earth, otherwise earth will become another Mars. If there are no greenhouse gases in the atmosphere, then there will be no life on Earth. Presence of greenhouse gases provides comfortable environment to entire life on Earth.

But emission of more greenhouse gases into the atmosphere gives adverse effects, it leads to global warming. Whereas absence of greenhouse gases in the atmosphere will decrease global temperature to less than 0°C. This phenomena is called as global dimming. Hence optimum level of greenhouse gases in the atmosphere is very much essential for providing comfortable climate for living organisms on the earth.

Prior to industrialization, CO_2 concentration in the atmosphere was 275 PPM. In that concentration, the Global average temperature was maintaining 15°C. At present CO_2 concentration in the atmosphere is more than 400 PPM. Hence the Global average temperature goes beyond 15°C. That is global warming. If the greenhouse gases rises further, the global average temperature also will rise proportional to it. In that same way, if the concentration of CO_2 in the atmosphere is brought down, the Global average temperature will decrease. If all greenhouse gases in the atmosphere are washed out completely, then the Global average temperature will be −19°C. In such situations, no life will be possible on earth.

Intensity of solar radiation and the distance between sun and earth are constant, even then the earth temperature is increasing continuously because of increasing trend in concentration of greenhouse gases in the earth atmosphere.

The planet's surface temperature is maintained by three factors viz solar intensity, distance between the sun and the planet and the greenhouse effect.

Solar radiation, CO_2 level and the average Global temperature can be compared with water pit which has a drainage hole and inlet pipe.

Imagine a pit with a inlet pipe which has constant flow of water. If drain hole is provided at the bottom of the pit, inlet water will be drained out from the pit immediately. There will be no accumulation of water in the pit, inlet Water will be drained out simultaneously. In that condition any living organism (fish or any other aquatic organisms) cannot live comfortably in that pit due to no water level. This condition may be compared with the atmosphere which has no greenhouse gases. (global temperature −19°C). If the drain hole is provided in the middle of the pit, then water level will be built up to the middle of the pit, only after that, water will start draining. Because of this sufficient water level, living organisms can thrive comfortably. This condition can be compared with greenhouse gases concentration (275 PPM) in the atmosphere at pre industrial period. (global average temperature +15°C). If no drain hole is provided in the pit then the level will build up to the top and over flows. In that condition, all the living organisms will be washed out. This condition can be compared with higher level of greenhouse gases concentration (high Global average temperature).

In all the three conditions inlet and outlet water quantity and flow are same, but water level is maintained as per the position of the drain hole. In the same way with respect to solar radiation input to the Earth and output from the Earth to sky are same at all the time irrespective of greenhouse gases concentration, but earth temperature will vary depending upon the greenhouse gases concentration.

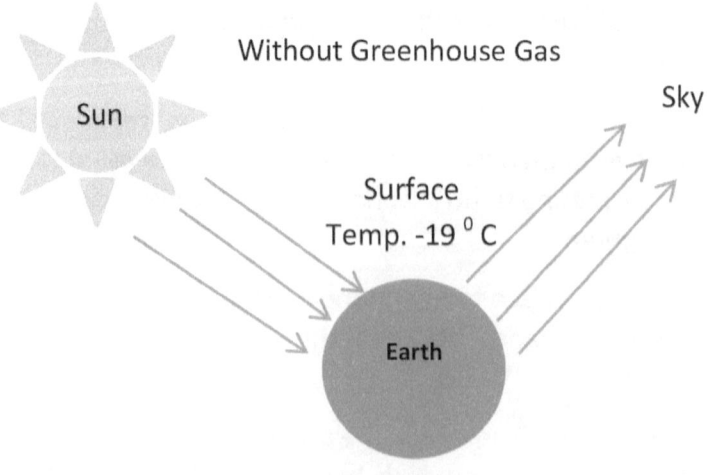

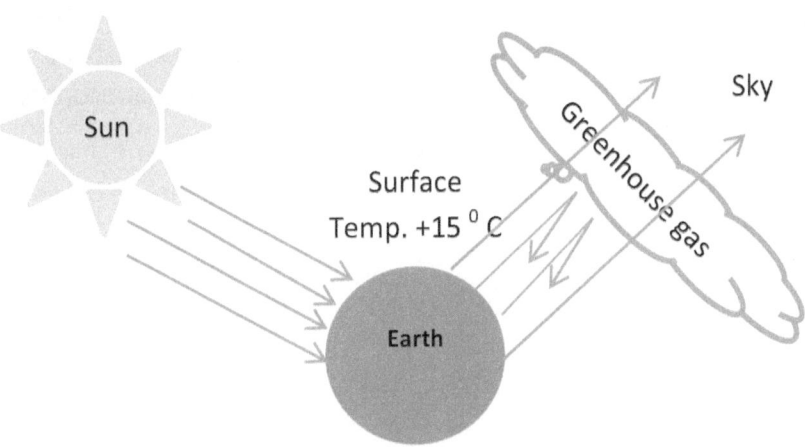

With Optimum level of
Greenhouse Gas

Hence optimum greenhouse gases concentration is required for comfortable life to all living organisms in earth. High CO_2 concentration in the atmosphere will wash out or burn out all living organisms due to unusual high global temperature. On the other hand, if there is zero CO_2 level in the atmosphere, there will be no chances of existence of life in earth due to very low temperatures (–19°C).

Comparison of Earth, Venus and Mars:

When we compare Earth, Mars and Venus, all the three planets are almost equal distance from the sun. Hence the solar intensities are constant to all three planets, but because of greenhouse gases concentration the surface temperature of these planets are getting varied

The atmosphere of Venus is composed primarily of carbon dioxide - 96.5% (atmospheric pressure of venus is 93 bar) The temperature of venus at the surface is 740 K.(465°C) because of more greenhouse effect. Hence venus is called as burning planet. In other hand mars has very little atmosphere. Its atmospheric pressure is only 0.007 bar. Because of this less greenhouse effect, Mars surface temperature is maintained very low at 222K (–53°C). It is called as frozen planet. Earth temperature is 288 K (15°C) because of amicable greenhouse effect. If the CO_2 concentration is increasing further in the earth atmosphere, then our Earth may also became an another Venus.

Even though the planet Mercury is very near to the Sun than the Venus, the surface temperature in Mercury is less than that of Venus because of less concentration of greenhouse gases in its atmosphere.

"You can make Mars as a living planet by warming it up with greenhouse gases, that is the positive effects of greenhouse gases"

12 CAN WE HOLD EARTH ON OUR HAND?!

yes! it is possible, thousands of hands joining together can hold the earth.

Everything in the universe is made up of atoms. Atoms have nucleus in the centre which consists of protons and neutrons. The nucleus possess more than 99.99% of the mass of the atom. The remaining mass of 0.01% is contributed by the electrons revolving around the nucleus. Normally the mass of electrons is not taken into account for any calculations since they have negligible mass. But when we compare the volume of nucleus and atom, the volume of nucleus is extremely minimum, which means more empty space is available between the nucleus and the outer most electron of an atom.

Here we can conclude mathematically that the entire thing in the universe including earth is mostly made up of empty space.

Atomic Volume:

The radius of an atom is 10^{-10} meter. (Say R)

The radius of nucleus is 10^{-15} meter. (say r)

(The average radius of atoms and nuclei are the value of 10^{-10} meter and 10^{-15} metre respectively irrespective of any atom from hydrogen to bigger atom like uranium.)

The volume of an atom:

$$= 4/3 \times \pi R^3$$

$$= 4/3 \times \pi (10 \times 10^{-10})^3$$

$$= 4/3 \times \pi (10 \times 10^{-30}) M^3$$

The volume of nucleus:

$$= 4/3 \times \pi r^3$$

$$= 4/3 \times \pi \, (10^{-15})^3$$

$$= 4/3 \times \pi \, (10^{-45}) \, M^3$$

When comparing the atomic volume to the nucleus volume, the nucleus volume is 10^{-15} times smaller than the atomic volume.

If we imagine the atomic volume of an atom as 10^{15} m^3, then the volume of nucleus of the same atom will be only one cubic meter. But the entire mass of an atom is contributed by the nucleus. 99.99% weight of an atom is only in the nucleus. In this way if we imagine an atom most of the space except the centre is empty.

Radius of an Atom and Nucleus

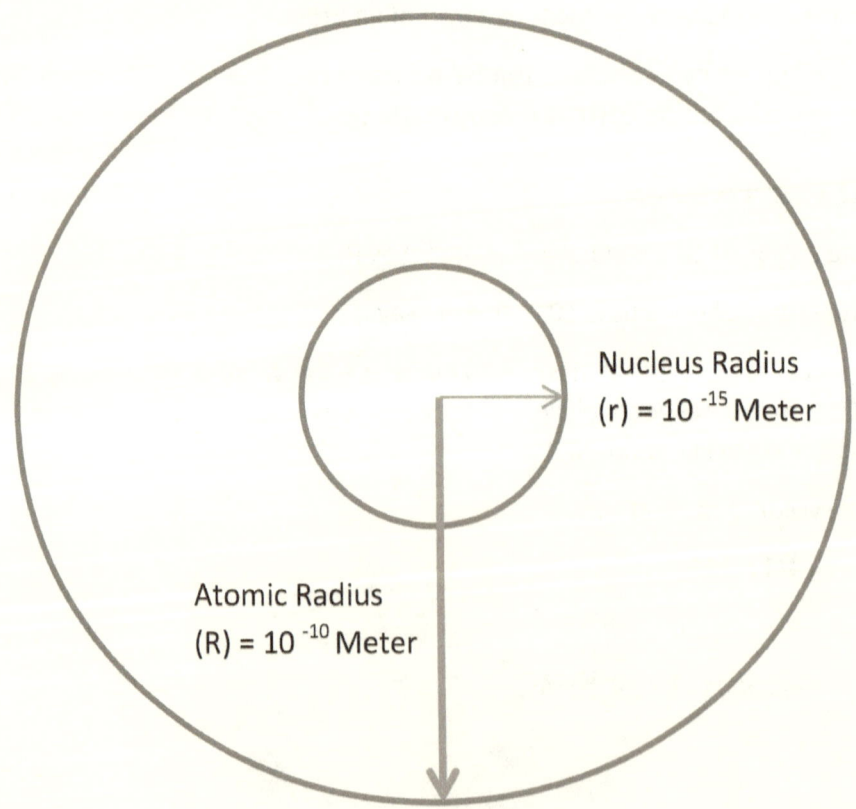

Nucleus Radius
(r) = 10^{-15} Meter

Atomic Radius
(R) = 10^{-10} Meter

How much space is empty?

Volume of Hydrogen atom:

A hydrogen atom has a single proton which is surrounded by a single electron.

The radius of hydrogen atom is equal to 0.529×10^{-10} meters.

It has a volume about 6.2×10^{-31} m^3.

Volume of Hydrogen Nucleus:

The protons have a radius of about 0.84×10^{-15} meters.

The volume of nucleus about 2.5×10^{-45} cubic meters.

Total volume of atom 6.2×10^{-31} m^3

Total volume of nucleus 2.5×10^{-45} m^3

Fraction of filled portion to the total volume $= \dfrac{2.5 \times 10^{-45} \text{ m}^3}{6.2 \times 10^{-31} \text{ m}^3}$

$= 4 \times 10^{-15}$

Percentage of filled volume to the total volume $= 4 \times 10^{-15} \times 100$

$= 4 \times 10^{-13}\%$

$= 0.0000000000004\%$

Therefore 0.0000000000004% of a hydrogen atom volume is filled with nucleus and rest of it is empty.

Percentage of Empty volume $= 100\% - 0.0000000000004\%$

Empty space $= 99.9999999999996\%$

A hydrogen atom has about 99.9999999999996% empty space. Similarly in any atom irrespective of the atomic weight, the empty space between the nucleus and the outer most electron is very high. Every object is made up of atoms and hence all objects are having more empty space. In the same way if the Earth is squeezed in order to collect all electrons to the centre, then the size of the earth would be a sphere having a diameter of about 200 meters.

Let us calculate the diameter of the earth mathematically if we shrunk it by eliminating all its empty space.

Mass of the Earth = 5.9722×10^{24} kg,

Nuclear Density:

Hydrogen has one proton in the centre

The protons have a radius of about = 0.84×10^{-15} meters.

The volume of hydrogen nucleus = 2.5×10^{-45} cubic meters.

Proton (nucleus) mass = 1.672×10^{-27} kgs

Hydrogen nucleus density = mass/volume

$$= \frac{1.672 \times 10^{-27} \text{ Kgs}}{2.5 \times 10^{-45} \text{ M}^3}$$

$$= 6 \times 10^{17} \text{ kgs/M}^3$$

Irrespective of the size of an atom the nucleus density is almost same for all elements, for example atomic weight of iron atom is 56. It is having 28 protons and 28 neutrons. Nucleus density of iron is calculated as follows:–

Mass of iron nucleus. = $56 \times 1.672 \times 10^{-27}$ kgs

Volume of iron nucleus = $56 \times 2.5 \times 10^{-45}$ M^3

Therefore iron nucleus density will be

$= 56 \times 1.672 \times 10^{-27}$ kgs/$56 \times 2.5 \times 10^{-45}$ M^3

$= 6 \times 10^{17}$ kgs/M^3 (Same as hydrogen nucleus density)

Mass of neutron is slightly higher than the mass of proton, hence there will be a small variation. However the average nuclear density is 2.3×10^{17} kg/m^3.

Mass of the Earth = 5.9722×10^{24} kg,

99.99% of Earth mass is contribution of the nucleons, that is protons and neutrons. It is connected with the nucleus density. Then the volume of the Earth will be (After deducting the empty space as occupied by all electrons)

$$= \frac{5.9722 \times 10^{24} \text{ Kg}}{2.3 \times 10^{17} \text{ Kg/m}^3}$$

$$= 2.5966087 \times 10^7 \text{ M}^3$$

$$= 25966087 \text{ m}^3$$

Volume of the earth $= 4/3\pi r^3$:

$$4/3\pi r^3 = 25966087 \text{ m}^3$$

Radius of the Earth (r) $= 183.674$ m

If we consider Earth sphere with all electrons brought to the centre together with the nucleus, then the radius of the Earth will be approximately only about 200 metres. Like this, a major portion of our body is considered to be of empty space. As per this theory, if our body is squeezed enough, then our body volume also will become one out of millionth part of a pin head.

Conclusion:

All the things in the universe including human body have more than 99.9% empty space. Ancient Tamil people used to tell as "உலகே மாயம்" and "காயமே இது பொய்யடா"

The exact meaning of this sentence is "The world is an illusion" and "The body is a lie"

"ANYTHING IS POSSIBLE IN THE UNIVERSE"

13 ONE OF THE SIDE EFFECT OF BURNING FOSSIL FUELS WILL BE COOLING OF THE EARTH'S SURFACE?

Yes, one of the side effect of burning fossil fuels is global dimming.

The main product of fossil fuel burning is carbon dioxide. It increases global warming. CO_2 and other greenhouse gases are blocking the outgoing radiation from the earth to space, causing global warming. On other hand, the byproduct of fossil fuel burning such as sulfur dioxide, soot and ashes are blocking incoming solar rays from the Sun and reflect them back to space thus preventing them from reaching the Earth surface, causing global dimming.

Let us see What is Global dimming?

The gradual reduction of solar radiation into the Earth's surface is called Global Dimming. It is believed that it has been caused by the increase in particulates such as sulphate aerosols carbon soot, ash in the atmosphere due to human activities, mainly due to burning of fossil fuels. Global dimming is opposite to Global warming, because it produces cooling effects. This dimming effect can be observed seasonally and in some specific areas with large numbers of industries.

Cooling effect caused by these pollutants may be comfortable to human life for a while, but they lead to highly undesirable ill effects and diseases to our respiratory system. These also causes thick fog, acid rain and air pollution.

Like global dimming there are some other interesting factors that give cooling effect to our earth. They are nuclear winter and contrails. what are nuclear winters and contrails and their cooling effects will be discussed here.

NUCLEAR WINTER

Nuclear winter is a severe and prolonged global climatic cooling effect that occurs after a nuclear war. The nuclear explosions would send massive clouds of dust into the stratosphere, blocking considerable amount of sunlight causing cooling effect. This phenomena is called as nuclear winter. The area of explosion would experience sudden decrease of temperature by 20°C for several months, and remain 2–6°C lower for 1–3 years. This cooling effect is because of blocking of natural solar incoming radiation that reaches the surface of the earth even upto 99%. This effect gets gradually cleared, over the course of several decades.

EFFECTS OF CONTRAILS

Also called as "condensation trails" or vapour trails, contrails are line shaped clouds produced by exhaust from aircraft engines. They also will produce the same cooling effect like aerosol and ash by blocking the incoming solar radiation. Quantifying this effect was extremely difficult because of continuous air traffic. But during Twin tower attack in USA on 11th september 2001,scientists got a rare opportunity to measure this effect. The US government had banned all the flights over its territory for three days after the attack. During this period, scientists observed an increase in temperature of 1°C (1.8°F) higher than the normal temperature.

> "If the greenhouse effect is a blanket in which we wrap ourselves to keep warm, nuclear winter kicks the blanket off"
>
> — Carl Sagan

14 COULD ICE CATCH FIRE?!

Yes! Methane clathrate is called as fire ice or methane ice. If we introduce a naked flame near methane clathrate, it ll start burning. Huge amount of fossil fuels discovered by humans is locked in the Arctic, Antarctic and permafrost areas in ice form. This vast quantity of natural resource of fossil fuel is called methane clathrate ($CH_4 \cdot 5.75H_2O$) or ($4CH_4 \cdot 23H_2O$) or methane hydrate. It is also called as hydro methane, methane ice, fire ice, natural gas hydrate, or gas hydrate. It is a solid compound found in the ice caps of north and south pole.

Methane hydrate is thought to be formed by the dead organic matter trapped as sediments under the ice sheets, millions of years ago like other fossil fuels. The trapped organic matter inside the ice were broken down by microorganisms anaerobically and produced methane. This process is similar to one that produces methane gas when the organic waste is dumped in the garbage pit and covered.

The following four places are identified as main sources of methane clathrate

1. Sediment and sedimentary rock units below Arctic permafrost

2. Sedimentary deposits along continental margins

3. Deep-water sediments of inland lakes and seas

4. Under the Antarctic ice core

Methane clathrate consists of methane, which is enclosed by frozen water. The methane molecule at the center will be completely surrounded by water molecules. Methane is trapped in the ice cubes, hence it is called as fire ice. If we introduce naked flame near methane clathrate, it will catch fire. Methane clathrate has very high concentration of methane. Melting of one cubic meter of methane hydrate yields about 160 cubic meters of gaseous methane.

Methane hydrate accumulations are found within a depth of few hundred metres from the surface.

But in antartic region they are found in greater depths.

Estimated Quantity of Methane Clathrate:

The total amount of carbon held by these sources are estimated approximately from 1000 to 5000 giga tonnes, which is anywhere between 100 to 500 times more than the quantity of carbon which are annually released into the atmosphere by the burning of fossil fuels. The quantity of methane in clathrate was estimated to be around 2×10^{16} m^3.

Methane has high calorific value than any other hydrocarbon or fossil fuel, including coal and gasoline. While burning, methane emits significantly very less quantity of CO_2 to the atmosphere than any other fuel. Methane can be taken from the clathrate by circulating hot water at a particular temperature in sustainable manner for power production. .

This methane clathrate will be stable as long as the temperature and pressure remains constant in the glacier region. Low temperature and high pressure are favourable conditions for methane clathrate to be in stable condition. High temperature and low pressure will stimulate methane clathrate and release huge amount of methane into the atmosphere. Methane clathrate can be compared with a sleeping tiger. As long as the tiger sleeps, it will not be a problem for anyone. Once it wakes up, then nobody can withstand its ferocity. When Methane clathrate is in stable condition, there won't be any harmful effects to the environment. If it is disturbed by global warming, then the ice surrounding the methane clathrate will melt and the methane molecules which were trapped inside the water molecules will be caged out, released to atmosphere and can lead to so many harmful effects, since it is a highly potent greenhouse gas than CO_2.

An additional problem due to release of methane into atmosphere is oxygen depletion in the sea water. More quantity of oxygen in seawater will be consumed when methane gets aerobic degradation and forms CO_2. The depletion of oxygen from oceans will have disastrous effects on marine life. Releasing increased amounts of CO_2 into the oceans will lead to acidification of water.

$$CH_4. +2O_2 \xrightarrow{\hspace{4cm}} CO_2 +2H_2O$$
(Aerobic degradation)

We have to give much attention to this highly potent energy source. If it is used properly with proper technology in a sustainable manner, desirable energy can be harvested from it. If proper care is not taken, then methane, being a highly potent greenhouse gas has the capability to destroy the entire environment.

> "No time before in human history has such a huge amount of ocean heat accumulated in the North Atlantic and North Pacific. This heat is now threatening to invade the Arctic Ocean and trigger huge temperature rises due to Methane eruptions from the Seafloor"
>
> — Sam Carana

15 IS THE DESERT TEMPERATURE AT NIGHT WARMER AS IT IS HIGH IN THE DAY?

No! the temperature in the deserts varies drastically from day to night. Deserts has a very hot temperature during day time and extremely cold temperature in nights. The average day temperature in the desert will be 38°C and in night, it will drop beyond (–) 40°C. In the day time sun radiation directly hits the desert surface and causes rise in temperature, whereas in night the sun no longer heats the desert and the heat from the day couldn't stay longer because of very little greenhouse effect.

All greenhouse gases except water vapour (moisture) spread uniformly all over the globe but the water vapour concentration will vary depending upon the areas like coastal, plains, forest and desert. Deserts are far away from the water bodies, hence the atmosphere above the deserts have very less moisture.

The areas having higher water vapour concentration (higher humidity) in atmosphere like coastal regions can trap the day time heat and store it for night time. But in desert areas due to low humidity, the heat cannot be trapped and held during nights. Hence all solar radiation fed in the day time would be completely escaped to the sky. Because of this, deserts get extremely cold at night. (Moisture is the major contributor of greenhouse effect than any other gas)

Like the deserts, in the planet Mercury, surface temperature extremely varies between day and night. Mercury is the first planet from the sun in the solar system. Due to this short distance mercury receives high intensity of solar radiation during day time.

The day time Mercury surface temperatures reaches up to 427°C. In contrast, night time the temperature drops to (–) 180°C, because of less greenhouse effect. The atmosphere of Mercury planet is only 10–14 bar, some billion times lesser than the earth atmosphere. Mercury atmosphere holds gases like helium, hydrogen, oxygen and water vapour in very minimum quantities as traces.

During night time, the atmosphere of mercury could not hold the heat received during the day. Therefore it experience very cold temperatures during night. The conditions existing in deserts are very similar to mercury and hence the temperature difference between day and night are also very similar to that planet.

> "In the empire of deserts water is the king and shadow is the queen"
>
> — Mehmet Murat ildan

16 CAN WE REDUCE GLOBAL WARMING BY ADDING IRON POWDER INTO THE OCEANS?!

Yes! Global warming can be reduced to an extent by the addition of small amount of iron powder into the oceans

Oceans play an important role in regulating the amount of CO_2 in the atmosphere because CO_2 can move quickly in and out of oceans. There exists an equilibrium with CO_2 concentration between atmosphere and the oceans. Whenever partial pressure of CO_2 exceeds in the atmosphere, the ocean absorb excess CO_2 from the atmosphere. In the same way when partial pressure of CO_2 increases in the ocean, it will be let out to atmosphere from the ocean. By this way CO_2 concentration is maintained in the atmosphere constantly.

Normally oceans emit CO_2 into the atmosphere, because oceans receive more carbon through the rivers and streams from the land biomass. The excess carbon received from the land (dry leaves, waste wood, dry grass, etc) is degraded into CO_2 and water, then CO_2 is sent to atmosphere from the oceans. This transfer of carbon between atmosphere and oceans maintain CO_2 levels in the atmosphere.

Approximately 90 giga ton to 100 giga tons of carbon moves forth and back between the atmosphere and the oceans annually.

CO_2 concentration had been maintained at 275 PPM for the past thousands of years till the start of industrial revolution period without much deviations. Atmospheric concentrations of CO_2 remains constant because the CO_2 being removed from the atmosphere will be exactly matched by the amount of CO_2 that is being added to the atmosphere.

Today, CO_2 concentrations in the atmosphere have been increasing as a direct result of human activities such as burning of fossil fuels, deforestation and other industrial activities. Over the past 150 years,

CO_2 concentration in the atmosphere have increased by as much as 40% (from 275 to 400 ppm). This excess CO_2 has to be accumulated either in the atmosphere or in the oceans. The CO_2 that remains in the atmosphere acts as a greenhouse gas and leads to global warming. But CO_2 taken up by the oceans does not affect the Earth's surface temperature.

The oceans contain about 50 times more CO_2 than the atmosphere. The oceans are able to hold more CO_2 than the atmosphere because most of the CO_2 that diffuses into the oceans reacts with the water to form carbonic acid and its dissociation products, bicarbonate and carbonate ions. Thus, CO_2 is converted from gaseous form into non gaseous forms such as carbonic acid bicarbonate and carbonate ions. This conversion effectively reduces the CO_2 gas partial pressure in the oceanic water, thereby allowing more diffusion of CO_2 from the atmosphere.

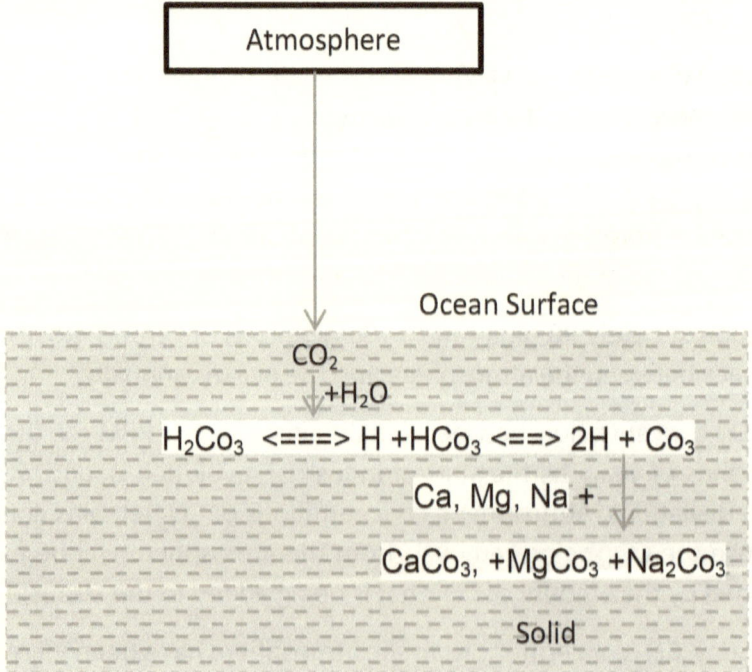

CO_2 react with water and form carbonic acid. This carbonic acid is dissociated to hydrogen and carbonate ions. These carbonate ions react with calcium, magnesium and other ions present in the ocean and form calcium carbonate and magnesium carbonate. These solid calcium and magnesium carbonate finally settle down in the bottom of the ocean floor. In this way ocean plays a critical role for capturing atmospheric CO_2 into the oceanic floor for long term storage.

Total quantity of fossil fuel usage around the world is approximately 7.1 giga tons per year. Out of this total quantity, approximately 2.6 giga ton of carbon is being taken by the oceans and 0.5 giga ton of carbon is taken by forest extra growth.(carbon fertilization effect). Every year approximately 3.5 giga tons of carbon is accumulated in the atmosphere as CO_2. This extra carbon accumulation in the atmosphere is the single main cause of global warming. If we increase the ability of oceans to recover more CO_2 from the atmosphere, then global warming can be reduced to a minimum level.

Marine plants and animals play a major role for the uptake of CO_2 in the oceans. Plants, primarily phytoplankton and also macro plants such as seaweed, take up CO_2 and release oxygen during photosynthesis.

No oceanic plants seem to grow faster in higher CO_2 environments, unlike many land plants. This is because the oceanic plants growth is generally limited by the availability of sunlight and the limitation of chemicals other than CO_2. Principally nitrogen, phosphorus, iron, zinc, and other micronutrients are limiting factors of the plant growth in the oceans. Among all these, iron is playing a major role in the phytoplankton (algae) growth.

Iron Fertilization Effect:

Iron fertilization is the intentional introduction of iron to iron-poor areas in the oceanic surface to stimulate phytoplankton growth. Therefore inexpensive small quantities of iron, can trigger large phytoplankton growth. The addition of one kg of iron powder into the ocean will stimulate the growth of 100,000 kg of plankton. The size of the iron particles is critical. Particles size in the range of 0.5–1 micrometer or less than 0.5 micron will be ideal for cyanobacteria and other phytoplankton to incorporate it in their biological activities and grow at a faster rate.

The resulting mass growth in population of sea plants will take more dissolved CO_2 into the oceans. Therefore more CO_2 will be diffused from the atmosphere into the oceans. Without healthy oceans, our life on Earth would be severely challenged. The oceans are considered to be the primary life support system of all living beings in the earth.

By this hypothesis, we can considerably reduce the global warming by the addition of small quantities of inexpensive iron powder into the ocean. In that way, the growth of oceanic plants would remove CO_2 from the atmosphere. The blooming plant biomass in the ocean can also be harvested and used as bio fuel, roof insulator and as cattle feed etc.

> "Never see what has been done,
> only see what remains to be done"

17 WILL THE GLOBAL TEMPERATURE DECREASE IF THE TOTAL SURFACE OF THE EARTH IS COVERED BY WATER?

No! In fact Global average temperature will increase.

The planet's surface temperature is primarily maintained by three factors.

1. Solar intensity
2. Greenhouse effects
3. Albedo

Solar Intensity:

Intensity of solar radiation depends on the distance between the sun and the planet. If the planet is close to the sun, intensity will be high and if it is away from the sun, intensity will be low. Our Earth is located 150 million kilometres away from the sun. The distance between the earth and the sun is constant. However, there will be slight variations in the distance during earth's orbiting around the sun. The nearest and farthest positions of earth to sun during its orbit are called Aphelion and Perihelion.

Perihelion position (147.1 million kilometers) is comparatively closer than the Aphelion (152.1 million kilometres), However these Perihelion and Aphelion effect do not make much difference in solar intensity on the earth. Hence solar radiation towards earth is constant. The sun supplies 342 joules of energy towards the earth per second per meter square.

Green House Effects:

The planet's surface temperature will also vary depending upon the concentration of greenhouse gases in the atmosphere. So increasing or decreasing the concentration of greenhouse gases change the planet's surface temperature. The greenhouse gases block the outgoing radiation from the planet towards outer space.

Albedo:

Albedo is the fraction of solar radiation reflected back to the sky from an object or surface of the Earth. It usually expressed in percentage. Albedo of an object is measured as ratio of reflected rays to the total input solar rays that fall on an object. For an example if the surface of an object reflects 30% of solar radiation, it means it absorbs 70% on its body. Then the albedo of that object is 30/100 (0.3) (Total solar radiation is 100%). Perfect white body does not absorb any heat on it, it reflects hundred percent of the radiation to the sky. Whereas the perfect black body will grasp all the radiation that falls on its surface. It will not reflect anything back to the sky. Hence the albedo of the perfect white body will be 1.0, and albedo of the perfect black body will be "0". However the albedo value changes based on the colour of the object.

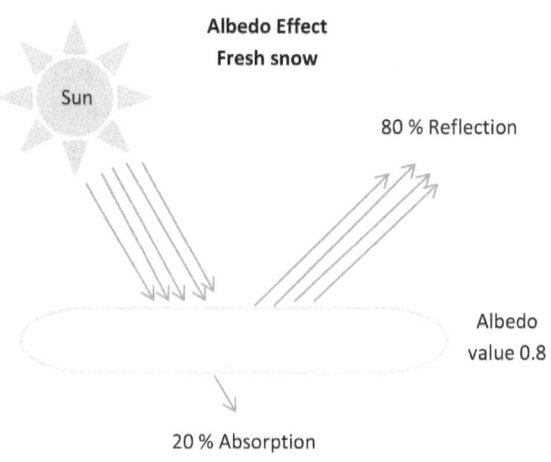

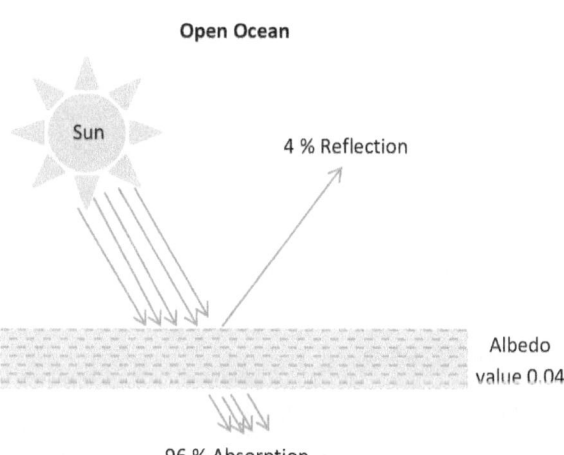

Albedo of the various surfaces are given below:

Surface	ALBEDO
Open ocean	0.04
Forest	0.08
Soil	0.17
Green grass	0.25
Desert sand	0.40
Ocean ice	0.5 to 0.7
Fresh snow	0.8

Earth is covered by variety of land and ocean surfaces like polar ice sheets, oceanic surfaces, snow, black soil, green vegetation, deserts etc. Each surface have different albedo value, but overall albedo of the Earth is 0.3. If the surface of the earth is covered more by snow and ice, then the average albedo will increase from its present value of 0.3. Similarly if the surface of the Earth is covered by more water, then the average albedo of the Earth will decrease to less than 0.3. Increasing albedo value leads to global dimming whereas decreasing albedo leads to global warming. Albedo of the fresh snow is 0.8 and albedo of ocean water is 0.04. If all the snow in the Earth's surface melts increasing the water surface area, then the average albedo of the Earth will decrease drastically.

So if the earth is completely covered by water, its average temperature will increase drastically because of low albedo. The albedo of water is very minimum compared to all other surfaces of the earth. If snow in the ice cap melts due to global warming then it will trigger further warming due to the phase changing of ice into water. So albedo of the planet play a significant roll in the surface temperature. Now the overall Earth albedo is 30% (0.30). If the entire surface of earth is covered with water, then the average albedo of the earth will become only 4% (0.04). In that condition, Earth will absorb 96% of solar radiation that falls on its surface and will reflect only 4% back to space.

Average albedo of the planets are given below:

Planets	ALBEDO
Mercury	0.06
Mars	0.29
Earth	0.30
Venus	0.75
Neptune	0.41
Uranus	0.41
Jupiter	0.41

Even though Venus has high albedo, its surface temperature is very high among all the planets, due to high greenhouse gases concentration in the atmosphere of Venus. To bring down earth's average temperature, we can follow Venus by increasing ice surface and we can follow Mars to bring down the greenhouse gases concentration. Ultimately more albedo and less greenhouse gases concentration will lead to less global temperature.

Conclusion:

As per the albedo concept, average temperature of earth will rise drastically if the Earth is completely covered by water. Hence the sea level rise not only will reduce the land area but will also increase the overall surface temperature of the Earth.

"The Earth is bathed in a flood of sunlight. If all this energy were captured by the Earth's atmosphere, its temperature would rise very fast. Luckily much of it radiates back to space. A good portion of Earth's albedo, or reflectivity, is created by its polar ice caps. If polar ice and snow were to shrink significantly, more solar energy would stay on Earth. Sunlight would penetrate oceans previously covered by ice, and warm the water. This would add heat and melt more ice, in a positive feedback loop."

– Kim Stanley Robinson

18 IS OZONE HOLE MERELY A SMALL HOLE?

No! Technically, the ozone hole is not a hole. The area in the atmosphere where the ozone layer is depleted below the minimum value is called as ozone hole. The size of the ozone hole is approximately 19 million square kilometers to 20 million square kilometers. This ozone hole appear in antartic region because of the special atmospheric and chemical conditions that exist there. The very low winter temperature in the Antarctic stratosphere causes the ozone hole.

Normally chlorofluorocarbon, a chemical used in refrigerators is the main reason for destruction of ozone layer. In the chlorofluorocarbon molecule only chlorine atom is the main cause for ozone depletion. The chlorine atom destroys the ozone molecule by the interaction of UV rays in the following manner.

UV rays split the Chlorofluorocarbons and releases chlorine atom as indicated in the following reaction.

$$CF_3CCl_3 \xrightarrow{\text{(UV rays)}} CF_3CCl_2 + Cl\cdot$$

The released chlorine atom will reacts with ozone molecule and destroys it as follows.

$$Cl\cdot + O_3 \xrightarrow{\text{(UV rays)}} ClO + O_2$$

This oxy chlorine reacts with another one ozone molecule and destroys it, again the chlorine atom is released.

$$ClO + O_3 \longrightarrow Cl\cdot + 2O_2$$

Like this one chlorine atom will stay more than 200 years in the atmosphere and destroy more than 100000 Ozone molecules. No chemicals in the stratosphere region is capable of destroying these stable chlorine atoms, however these chlorine atoms will get converted by the water and No3 molecules available in the atmosphere in the following manner.

$$Cl\bullet + H_2O \longrightarrow HCl + O_2$$

$$Cl\bullet + NO_3 \longrightarrow ClONO_2 \text{ (chlorine nitrate)}$$

These compounds (HCL and $ClONO_3$) will stay in the stratospheric region in stable condition. They cannot be disturbed or destroyed by any means. This chlorine atom will not be released from the two molecules (HCL and $ClONO_2$) under normal conditions. Hence large quantity of HCL and chlorino nitrate molecules gets accumulated in the stratosphere. They will be activated in the very low Antarctic temperature and releases free chlorine atoms. When Antarctic spring arrives in September. Stratosphere temperature will be decreased below $-90°C$. This very cold temperature is enough to form polar stratospheric cloud. This cloud provides active surface for the HCL and $ClONO_3$ and the following reactions will happen in that extreme cold temperature and sets the chlorine atom free.

$$ClONO_2 + H_2O \longrightarrow HOCl + HNO_3$$

$$HOCl + HCl \longrightarrow Cl_2 + H_2O$$

$$ClONO_2 + HCl \longrightarrow Cl_2 + HNO_3$$

As long as the spring season exists, these released chlorine molecules will be accumulated in huge quantity in that region and stay without any action until the sun rises. Once the Sun rises, the chlorine molecules will be split by the UV rays and releases chlorine atoms. These huge quantity of released chlorine atoms will destroy ozone molecules more vigorously and as a result, overall ozone concentration will decline in that area.

$$Cl_2 \xrightarrow{\text{(UV rays)}} 2Cl$$

$$Cl\bullet + O_3 \xrightarrow{\text{(UV rays)}} ClO + O_2$$

In this way the ozone concentration gets depleted in the atmosphere over Antarctica and creates ozone hole. After the season changes to summer, become normal, ozone molecules in other parts of atmosphere will rush up to Antarctica and normalises the ozone concentration.

However the overall ozone concentration in the atmosphere gets depleted in this manner.

Bromine is also a very potent ozone depleting gas than chlorine. The source of bromine gas is methyl Bromide (CH_3 Br) which is used in agriculture for sterilising the soil.

For a period of two to three months, approximately 50% of the total amount of ozone in the stratosphere above Antartica disappears. At some extreme condition it even increases to as much as 90%. This is called as the Antarctic ozone hole. After the spring season, temperature begin to rise, the ice evaporates, and the ozone layer starts to recover.

The ozone concentration is measured in Dobson units (DU). The normal ozone concentration in the atmosphere is 350 dobson units. Which means if all the ozone in the atmosphere is compressed at sea level around the globe in normal temperature and pressure the size of ozone layer will be 3.5 mm. 1 Dobson unit is equivalent to a layer of ozone 0.01 mm thickness at one atmospheric pressure and 0°C. During the annual appearance of Ozone hole in the Antarctica, it will be well below 100 DU.

Now alternative chemicals like Hydro chloroflurocarbon (HCFC) and Hydro flurocarbon (HFC) are used in place of chlorofluorocarbon in refrigerators which reduced the ozone depletion considerably and improved the ozone concentration in the stratosphere region. These Hydro chloroflurocarbon (HCFC) and Hydro flurocarbon (HFC) are more active than the CFC, hence it will be destroyed before reaching the stratosphere region.

"Earth without OZONE is like a house without roof"

SAVE OZONE

19 HOW MUCH CARBON DIOXIDE CONCENTRATION IN THE ATMOSPHERE IS EXPECTED TO RISE IF ALL THE FOSSILS ARE BURNT OUT?

Total Carbon Storage in the Globe:

Total quantity of carbon present in the globe is constant, only a small variation will be there due to the addition of C-14 (isotope of carbon), which are formed from nitrogen atoms in nature by the interaction of cosmic rays.

Carbon atoms may shift from one storage to another storage on its cycle, for example carbon atom present in the atmosphere as CO_2 will transfer to plants in glucose molecules via photosynthesis. In the same way mineral carbonate on erosion releases CO_2 into the atmosphere but overall quantity of carbon in the globe will not vary.

Carbon Distribution:

All over the globe, Carbon is present either in inorganic or organic form. Inorganic carbon is carbon extracted from ores and minerals. Some examples of inorganic carbons are carbon monoxide, carbon dioxide, polyatomic ions, cyanide, cyanate, thiocyanate, carbonate and carbide etc. Organic carbon is derived from living organisms via photosynthesis. Various storages of carbon and its quantity in each area are described here.

Lithosphere:

In the interior of the earth the total carbon present is estimated at 15 000000 Giga tons (Giga ton $=10^9$ ton). Mostly these carbon are inorganic and in the form of mineral carbonates and bicarbonates.

(calcium, magnesium and sodium carbonates). Minimum quantity of organic carbon is available as fossil fuels in the interior of the lithosphere.

Fossil Fuels:

Fossil fuels are ancient carbon produced by ancient photosynthesis, which are stored under ground in the form of coal, oil, natural gas, methane clathrate and peat. The total quantity of fossil fuels are estimated as 4130 giga tons. Individual quantities of fossil fuels in each forms are given below.

Coal	3510 Gt
Oil	230 Gt
Natural gas	140 Gt
Peat	250 Gt

(Peat is partially decayed vegetation or organic matter. It is locked in wet areas called Peat land)

Soil Carbon (1550 Gt)

All around the globe, at a depth of approximately 1 metre, we can find plant litter and humes (Plant waste on the ground). These are organic carbon. Mostly when plants die or harvested, the bottom portion of the plants (roots) remains in the soil. This soil carbon quantity is more than the total quantity of carbon available in atmosphere and biosphere combined.

Methane Clathrate ($CH_4 \cdot 5.75H_2O$) (2000 to 5000 Gt)

Methane clathrate is also a fossil fuel that are available beneath polar ice sheets. These carbon storages are unidentified and are a huge source of fossil fuel. These methane clathrates were formed when huge ancient plants were masked by the Arctic and Antarctic ice sheets. Huge quantity of methane gas was formed from these locked biomass on anaerobic degradation and are available beneath the polar ice sheets. The locked up biomass Methane molecules are caged by the surrounding water molecules in the stable crystal form. It is also called methane hydrate, hydro methane, methane ice, fire ice, natural gas hydrate, or gas hydrate.

Oceanic Carbon 38400 Gt

Oceanic carbons mostly are inorganic as in the form of carbonate, bicarbonate ions and CO_2 gas dissolved in the oceanic water. Out of total carbon stored in the ocean, 1000 Giga tons are organic carbon present mostly in the form of bio organisms.

Carbon in biosphere (602Gt)

Land biomass 600 GT

Ocean biomass 2 GT

(All the biomass of plants and animals)

Atmospheric carbon (720 GT)

In atmosphere, carbon is stored as carbondioxide, carbon monoxide, methane and other traces of hydrocarbons. But major part is occupied by CO_2

Plants trap atmospheric carbon in its body as in the form of carbohydrates by the process of photosynthesis. And we know that the ancient plants and animals bodies were converted into fossil fuel, soil carbon, oceanic carbon, methane clathrate etc. The carbon present in these fossil fuels must have been once existed as atmospheric carbon in earlier periods.

Let us calculate the total increase in carbon dioxide concentration in the atmosphere, if we burn all the fossil fuels within this century in unsustainable manner.

Total quantity organic carbon available at present:

Fossil fuels	4130 Gt
Oceanic organic carbon	1000 Gt
Soil Organic carbon	1550 Gt
Carbon in biosphere	600 Gt
Methane clathrate (approximately)	2000 Gt

(Total quantity of organic carbon 9280 GT)

The mass of the atmosphere 5.12×10^{21} grams

Average density of the air 1291.4 gram/ m^3.

Total volume of air in the atmosphere = mass/density

$$= \frac{5.12 \times 10^{21} \text{ g}}{1291.4 \text{ g/m}^3}$$

Total volume = $3.96469 \times 10^{18} m^3$

The quantity of CO_2 required to raise 1vpm in the atmosphere

$$= \frac{1}{10^6} 3.96 \times 10^{18}$$

$$= 3.96 \times 10^{12} m^3 \text{ of } CO_2$$

Mass of the CO_2 required for raising 1 ppm in the atmosphere

$$= 3.96 \times 10^{12} \times 1.964$$

(1m^3 of CO_2 mass 1.964 kg)

$$= 7.777 \times 10^{12} \text{ kgs}$$

Mass of carbon required for rising 1ppm in the atmosphere.

(12 gram of carbon available in the 44 gram of CO_2)

$$= \frac{12}{44} \times 7.77 \times 10^{12}$$

$$= 2.12 \times 10^{12} \text{ kgs}$$

$$= 2.12 \times 10^9 \text{ ton}$$

$$= 2.12 \text{ Gt.}$$

2.12 GT of carbon is required to raise 1ppmv of CO_2 concentration in the atmosphere.

Total organic carbon available as fossil fuels and other sources = 9280 GT.

(This quantity of carbon was originally obtained only from atmosphere) if we burn all the fossil fuels within this century in unsustainable manner

If we burn all the fossil fuels available in the earth completely within this century in unsustainable manner then the future concentration of CO_2 in the atmosphere will be 4363 ppm

$$= \frac{1}{2.12} \times 9280$$

$$= 4363 \text{ ppm}$$

If we add present CO_2 concentration in the atmosphere into the calculated value

$$= 4363 + 400$$

$$= 4763 \text{ ppm.}$$

During pre industrialisation era, the value of CO_2 concentration was maintained constantly at 275 ppm. At present the CO_2 concentration in the atmosphere is more than 400 ppm. That means after industrialisation, extra CO_2 had accumulated in the atmosphere due to the excess burning of fossil fuels. If we burn all the available fossil fuels completely within this century in a unsustainable manner then the CO_2 concentration in the atmosphere will rise an extra 4363 ppm in addition to the present value of 400 ppm. Hence the expected CO_2 concentration in the atmosphere will be more than 4763ppm.

All things in the universe will proceed from orderliness to disorderliness naturally. By that way, naturally all the organic carbon available in all the spheres tend to move towards atmosphere as CO_2. But that natural process will take very long time, and that duration will be enough for oceans to absorb excess CO_2 from the atmosphere. Due to human greed and their need for instant gratification, we burn fossil fuels in very large amounts beyond the capacities of natural carbon sink. Carbon sink denotes natural reservoirs like oceans that absorbs and stores the atmosphere's carbon with physical and biological mechanisms. If we are utilizing the fossil fuels slowly in a sustainable manner, then the oceans and forests will take care of everything and control the CO_2 concentration in the atmosphere.

Our aim should be to increase organic carbon by increasing the capacity of natural carbon reservoirs such as forests and biodiversity of ocean and land.

"We are running the most dangerous experiment in history right now, which is to see how much CO_2 the atmosphere can handle before there is an environmental catastrophe."

– Elon Musk

20 ARE THERE ANY NATURAL NUCLEAR REACTORS IN THE INTERIOR OF THE EARTH?

Yes! Nuclear fission reaction is happening in the interior of the earth similar to man-made nuclear reactors. Earth's surface receives heat energy not only from solar radiation, but also from the interior of the earth, more specifically from the core. This heat energy that comes from the interior of the earth is estimated to be 0.025% of earth's total energy budget.

Total energy to the earth's surface is calculated as mentioned below:

Solar energy   99.97% (173,000 Tera watts or 173000×10^{12} watts)

Geo thermal energy   0.025% (44.0 Tera watts)

Tide energy   0.005%

Energy from sources other than the solar system. Traces

Earth surface continuously receives energy from the above mentioned sources. Among all the four sources, the dominant energy source is the solar energy. Another significant source is from inner core of earth called Geo thermal energy. Huge amount of heat energy reaches to the earth surface from the inner core of the Earth. Hot springs and volcanoes are the examples of heat energy obtained from the interior of earth. The flow of heat from Earth's interior to the surface is estimated at 44×10^{12} watts. (44×10^{12} joule/second).

The two source of Geothermal energy:

1. Ancient heat available during the formation of the Earth

2. Nuclear fission energy

Ancient Heat Available During the Formation of Earth

The Heat developed during the earth formation is locked in the interior of Earth and is roughly estimated as 22 Tera (22×10^{12}) watts. The geothermal energy of earth's crust originated during the time of formation of the planet.

Nuclear Fission Energy

Earth's interior has many radioactive elements—primarily uranium, thorium and potassium. Over billions of years, the radioactive isotopes have been splitting and releasing energy as well as antineutrinos, just like man-made nuclear reactors. This energy is coming out to the outer surface of the earth instantly. By measuring the antineutrino emissions, scientists can determine how much of Earth's heat results from this radioactive decay.

Fission reactors beneath the Earth might have been burning for billions of years.

Uranium could become sufficiently concentrated at the base of Earth's mantle to ignite self-sustained nuclear fission, as in a man-made reactor. Radioactive decay of unstable isotopes of heavy elements such as uranium happens all the time beneath the Earth's surface. The uranium and thorium accumulated in the Earth's crust and mantle are estimated to be 50,000 and 160,000 billion tons respectively.

Four radioactive isotopes are responsible for the majority of radiogenic heat because their enricment is relatively higher to other radioactive elements.

Radioactive Isotopes in the Earth Crust Are as Follows:

Uranium-238 (238U)

Uranium-235 (235U)

Thorium-232 (232Th)

Potassium-40 (40K)

In this, potassium decay is estimated as 4.0 tera watts. This Geo thermal energy can also be harvested for production of electric power. Many countries have already installed geothermal power plants.

Earth is receiving nuclear fusion energy from the sun. It also receives nuclear fission energy from the inner core of the earth. The main source of energy for the earth is nuclear energy. But people in some parts of India are quiet agitated about the installation of nuclear reactors. There is a huge nuclear fusion plant at the top of the Earth (Sun) and also nuclear fission plants at the bottom of the Earth . Then why not in the middle with a safe and fool proof system?.

"THE INTERIOR OF THE EARTH IS EXTREMELY HOT –SEVERAL MILLION DEGREES

IT IS ONLY BECAUSE OF NATURAL REACTORS"

21 IS THE DURATION OF A DAY IS MORE THAN THE DURATION OF A YEAR IN VENUS? HOW?

Yes, It is correct. The time taken by a planet for one full rotation on its own axis is called a day. Whereas the time taken by a planet to go around the sun in one revolution is called a year. The planet Venus is taking more time to rotate on its own axis than the time it takes to go around the sun.

VENUS

Venus looks like a very active planet. It has mountains and volcanoes. Venus is similar in size to Earth. Earth is just a little bit bigger than venus. Venus is the second planet from the Sun. Venus is the only planet in the Solar System where days are longer than years. Its rotation on its own axis is very slow. It takes about 243 Earth days for one full rotation. Since Venus is very close to the sun than the earth, it takes just 225 days for a full revolution around the sun. That is why, one day on Venus is 18 earth days more than one year. Days of venus are the longest in solar system. In earth Sun rises and sets once in each day, that is once in 24 hours. But on Venus. sun rises once in every 117 Earth days.

Another interesting thing to note is Venus rotates in opposite direction (anticlockwise) to that of the Earth and other planets (all of which rotate in clockwise direction). It is believed that, millions of years ago, when the Solar System was in the process of formation, Venus was hit by a huge asteroid, which altered its axial rotation and its direction.

"Our life span is only 150 days in Venus"

FINAL CONCLUSION

எவ்வ துறைவது உலகம் உலகத்தோடு
அவ்வ துறைவ தறிவு.

As dwells the world, so with the world to dwell
In harmony – this is to wisely live.

உலகம் எவ்வாறு நடைபெறுகின்றதோ, உலகத்தோடு பொருந்திய வகையில் தானும் அவ்வாறு நடப்பதே அறிவாகும்.உலகத்தில் நடைபெறும் இயற்கை நிகழ்வுகள், அவற்றின் தன்மை, அந்த இயற்கையான நிகழ்வுகள் நமது செயல்களால் பாதிக்கப்பட்டால் வரும் தீமை இவைகளை முறையாக அறிந்துகொண்டு நாம் நமது அனைத்து செயல்களையும் முறைப்படுத்தவேண்டும்.

ஒவ்வொரு வினாடியும் உலகம் தன் இயற்கை நிகழ்வுகளால் தன்னைப் புதுப்பித்துக்கொண்டிருக்கிறது. வாயுமண்டலமும் நீர்மண்டலமும் எப்பொழுதும் சுழற்சியில் இருந்துகொண்டே அனைத்து உயிர்க்கோலத்தின் இயக்கங்களை நெறிமுறைப்படுத்திக் கொண்டிருக்கின்றன. நம்முடைய ஒவ்வொரு நிகழ்வுகளும் ஒவ்வொரு கண்டுபிடிப்புக்களும் உலகத்தின் இயற்கைத் தன்மையையும் அதன் சுழற்சியையும் பாதுகாக்கும்வண்ணம் இருக்கவேண்டும்

It is always wise to live along with the world on its terms. We should regulate our activities in concordance with nature. We should foresee the adverse effects on nature if it is disturbed and correct the same to ensure a safe environment.

Earth is continuously working as a heat engine for the benefit of the biosphere. Biosphere means the part of the earth's surface, ranging from a depth of 6000 meter deep in sea and upto 2000 meter height in atmosphere where all organisms exist. All natural processes in the earth are happening to bring evenness from unevenness.

For example, composition of air in all altitude around the globe is constant. Even if we pollute a particular area by venting any hazardous gas, immediately it will be diluted by means of dispersion and spreading all over the globe. Otherwise the polluted area becomes more harmful and cause severe damage to the people living there. It is being done by the movement of wind. Wind movement not only bring evenness in composition of air in the atmosphere. It also attempts to maintain uniform temperature all around the globe.

Likewise, through hydrological cycle, water is being circulated globally. Even desert areas where no water bodies are available gets some quantity of rain due to this water cycle and helps the living organisms in that area.

Like air cycle, water cycle also dissipates the local pollution levels. So many Biogeochemical cycles such as carbon cycle, oxygen cycle, nitrogen cycle, phosphorus cycle and sulfur cycle are happening continuously for distribution of all essential nutrients worldwide.

Frozen permafrost soil is the perfect place for bacteria and virus to remain alive in a dormant state for very long periods of time. Significant numbers of viable ancient microorganisms are known to be present within the permafrost. They have been found in both polar regions up to 400 m depth in ground and at very low temperatures of −27°C. As the Earth warms, more permafrost gets melt, gradually exposing them. It means melting ice could open the door for ancient bacteria and viruses. These pathogenic viruses and bacterias can infect humans and animals, some of them might even cause a global pandemic, against which we might have no antidote or vaccine. The temperature in the Arctic Circle is rising quickly, about three times faster than the rest of the world. Hence there will be a big threat on all living things in future.

In all our inventions and technological developments we shall aim to maintain the temperature of earth surface as well as ocean to ensure that temperature is not getting increased. Atmospheric temperature rise will trigger global warming, polar ice melting and destruction of biodiversity. Oceanic temperature will raise the total sea heat content, which will drive out oxygen and CO_2 gases from the sea to atmosphere as solubilities of both the gases are low at higher temperature. Depletion of CO_2 and oxygen will reduce the biological activities in the ocean and at the same time, the driven out CO_2 in to atmosphere will enhance greenhouse effect.

Hence all our activities must be aligned with earth's natural processes. We should not deviate and disturb the system, in order to ensure a safe and healthy environment.

Ultimately, If we can ensure a stable global temperature through sustainable developments, the world will march towards stable future for our generations.

ABOUT THE AUTHOR

Hailing from a remote village named Perichampakkam, in Cuddalore district, the author, S.Kalaivanan, belongs to an agriculture family.

After graduating in Chemistry from the Government Arts college, Villupuram, he started his career in Heavy water plant, (Department of Atomic energy) Tuticorin and is still working there, building a reputation of a straightforward and honest person with an unblemished career spanning 36 years.

Ever penchant for knowledge, he completed his Master's in Environmental Science in Manomaniam Sundaranar University, Tirunelveli.

Kalaivanan is very interested in Tamil culture and the way of life of ancient Tamil people and has read a multitude of Tamil literature books especially *Thirukkural* and *Naladiyar*.

In the *Thirukkural,* Thiruvalluvar covered all the areas of society and how to lead a life for every individual—son, father, mother, wife, husband, friend, servant, soldier, minister and king. He also touched environmental topics such as biodiversity, atmosphere, water, soil, river ocean etc.

The author is someone who strongly believes that everyone deserves an equal share of resources. He has a tendency of helping people without advertising, and was inspired by *Thirukkural* and wrote this book about the present environmental issues and presents solutions based on it.

Being a lover of Mathematics, in this book he made his own calculations for predicting natural process.

www.ingramcontent.com/pod-product-compliance
Lightning Source LLC
Chambersburg PA
CBHW021404210526
45463CB00001B/215

* 9 7 8 1 6 4 8 9 2 6 6 3 1 *